Fred
Thanks for trimming the hair out of my eyes before these trips

Marine Life of Johnston Atoll, Central Pacific Ocean

Phillip S. Lobel

Natural World Press, Inc.

Printed in the United States of America.

ISBN: 0-939560-27-5

Author's Note

The research for this book was supported by funding from the Department of Defense in support of the missions by:

The Program Manager for Chemical Demilitarization, US Army,

The US Air Force,

The Defense Threat Reduction Agency,

The US Navy,

The Legacy Resource Management Program, and

The US Coast Guard

Research was initiated when I was a Postdoctoral Fellow in Oceanography at the Center for Earth and Planetary Physics, Harvard University, continued while I was on the scientific staff of the Woods Hole Oceanographic Institution and was finally completed in my position as a professor in the Boston Univeristy Marine Program in Woods Hole, MA

The majority of photographs in this book were taken at Johnston Atoll. However, some photos were taken at other locations of the same species that occur on Johnston Atoll but for which the author did not have quality photographs from there. Supplementary photos were taken in Hawaii (n= 33), Fanning Atoll (n= 10), and Wake Atoll (n=2) and Palau (n=1).

Contents

Introduction

The purpose of this book is to provide an identification guide for the fishes and other common marine life of Johnston Atoll (JA), Central Pacific Ocean (16° 45'N, 169° 31'W). These reefs and fishes are currently under study to determine whether some 70 years of military activities on the atoll have had measurable and significant ecological impacts. Presently the atoll is jointly managed by the US Army, Air Force and the US Fish and Wildlife Service (USFWS) as a military base and a National Wildlife Refuge . This guide illustrates the majority of fishes and the most abundant corals, echinoderms and other marine species that one normally encounters on JA reefs.

This book is produced with funding from the Department of Defense Legacy Resource Management Program and the US Army Program for Chemical Demilitarization for the purposes of educating Administrators, Political leaders, Island workers and the American public about the ecological beauty and relatively pristine reefs that abound at Johnston Atoll. This ecosystem is maintained and thrives under the combined stewardship of the US Department of Defense and the USFWS National Wildlife Refuge system. For more information on the Department of Defense Coral Reef Protection Plan see the following websites on the DoD website "DENIX"

The Coral Reef Conservation Guide for the Military
http://osiris.cso.uiuc.edu/denix/Public/ES-Programs/Conservation/Legacy/Coral/coral.html
and *The Coral Reef Protection Implementation Plan*
http://www.denix.osd.mil/denix/Public/ES-Programs/Conservation/Legacy/natural_water.html

This book provides a reference guide to the identification of species that have been the subject of many ecological and biodiversity studies on Johnston Atoll (see bibliography). A majority of these studies have been conducted in order to help evaluate whether or not military activities have contaminated the reefs. A complete listing of all fishes can be found in Randall et al 1985 and Kosaki et al 1991. A detailed terrestrial inventory, history and ecological review are in Amerson and Shelton 1976. The reader interested in learning more about Hawaiian Reef Fishes (and their Hawaiian names) is referred to Dr. John E. Randall's book, *Shore Fishes of Hawaii,* for further information. Since many of the same species occur at Johnston Atoll, Dr. Randall's reference is a necessary companion guide to this book.

Johnston Atoll is a truly remarkable place. The Atoll is an unincorporated territory of the United States of America currently used as a secure military installation isolated in the Central Pacific Ocean at 16o 45' N, 169 o 31' W. Honolulu, Hawaii is about 800 miles (1,287 km) to the northeast. The nearest landfall is French Frigate Shoals about 500 miles (804 km) north in the Northwest Hawaiian Islands. The Line Islands of Kiribati are about 900 miles (1,448 km) south and the Marshall Islands are 1600 miles (2,575 km) to the southwest.

This remote atoll is completely isolated in the middle of the deep blue ocean and is home to both high tech military operations and beautiful coral reef habitat. It has an obscure history as far as the general public is aware but one that is well known to the tens of thousands of Americans who have been stationed there over the past 70 years. As I work in our east coast home on Cape Cod, our neighbor a few doors up the street, now in his 70's, was once at Johnston in transit on a military aircraft during WWII The late postmaster at our local Post office in Woods Hole was a sergeant in the USAF during the 1960's nuclear testing era. Another local friend of mine refueled his fighter plane there enroute to the Korean War. I have met dozens of people worldwide who have been on Johnston Island at one time or another. It is amazing to consider, that even though Johnston Atoll is such a small place on a relative scale, the incredible number of people connected with the island and the rather large changes that have occurred there over the years.

The military mission on Johnston Atoll overlaps with it also being a remote National Wildlife Refuge managed by the US Fish and Wildlife Service. The USFWS activities have focused primarily on the protection of migratory seabirds. The atoll has a long military history with all sorts of activities. It has been used for nuclear bomb testing, storage of the defoliant "Agent Orange", as a transit airfield for military planes, as a submarine base, a Loran station and as a storage site for chemical weapons. Beginning in 1990, the main mission on the atoll has been the Army's chemical weapon demilitarization project or JACADS for the Johnston Atoll Chemical Agent Disposal System. This project will result in destruction of 6.6% of the nation's military stockpile of chemical weapons. These weapons include the deadliest chemicals ever used in bombs, mines and rockets including different types of nerve and mustard agents.. After ten years of operation, JACADS completed its mission early in 2001. What will happen on the atoll next has as yet to be determined, but its role as a National Wildlife Refuge is becoming increasingly important. Conservation of the marine resources of the atoll are especially in focus at the national level following the 1998 Presidential Executive Order 13089 which provided for the protection of coral reef ecosystems. Despite its military occupation, the atoll's underwater resources are in nearly pristine condition. Big fishes are abundant in numbers that are rarely seen on other Pacific reefs, where high levels of fishing pressure occur due to the usual human population .

A Brief History

Johnston Atoll was first discovered on 2 September 1796 by a whaling ship, the brig *Sally* from Boston, Massachusetts. However, Capt. Charles Johnston of the British ship HMS Cornwallis formally documented the island on 14 December 1807 and it is for him that the atoll and the main island are named. Therefore, it is not, perhaps, so ironic that the marine ecological research and monitoring program is mostly conducted by scientists based out of Woods Hole, Massachusetts. It is a further small-world kind of coincidence that the whaling ship, Howland, which shipwrecked on the atoll on 26 December 1889, was also from New Bedford, Massachusetts which is just across the bay from where I now live on Cape Cod.

Johnston Atoll has an unusual geography, which is unlike most atolls that have barrier reefs all around. JA has a semicircular emergent barrier reef only around the northern and western margins of the atoll platform. Reefs to the east and around the south are submerged from a few to several meters. The prevailing trade winds blow in from the east and, without an emergent reef barrier, drive current flow across the atoll.

The age of Johnston Atoll has been estimated at around 80 million years old, making it one of the oldest atolls in the world. Originally, there were two islands; the largest, Johnston Island, was about 46 acres (0.18 km2) and the smaller, Sand Island, was about 10 acres (0.04 km2) in area. The islands were formed as mounds of guano were deposited on dead coral and sand that built up in shallow water. Johnston Atoll has no natural source of freshwater and humans including Pacific Islanders have never inhabited the atoll nor has evidence ever been found of long-term visitations prior to the 1800s.

Today, after years of American engineering, the original islands have been enlarged dramatically and two additional islands were created by dredge and fill operations. The intensity of military activity, and therefore the need for more space on the atoll began before WWII and then increased in the 1960's with the start of the US nuclear testing program in the Pacific. Johnston Atoll was the site of a series of high altitude nuclear tests. Since WWII, Sand Island was increased from about 10 to 22 acres (0.04 to 0.09 km2) and Johnston Island was increased from 46 to about 640 acres (0.18 to 2.58 km2). Two new islands,

North and East Islands were artificially created by dredging coral from the lagoon and are 25 and 18 acres (0.10 and 0.07 km2) in size, respectively. Overall the lagoon covers and area of about 39,000 acres (158 km2).

Ownership of the atoll was established with the passing of the US Guano Act that guaranteed military protection to citizens' establishing guano-mining operations on uninhabited and unclaimed guano-laden islands. Americans seeking their fortune in guano collecting claimed Johnston Atoll on 19 March 1858. However, the Kingdom of Hawaii contested this claim on 27 July 1858 whcn King Kamehaeha proclaimed the annexation of Johnston Atoll, which he called Kalama Atoll. Once the US annexed Hawaii in 1898, it gained an undisputed claim to the atoll.

The first scientific survey to JA was led by Dr. Alexander Wetmore from the Bishop Museum, Oahu, HI, in 1923, with support from the Department of Agriculture and the US Navy. They arrived by boat and landed on the largest island of the atoll, Johnston Island on 10 July 1923. The survey report described the incredible diversity of seabirds, fishes and other marine life abundantly flourishing at Johnston Atoll. The result was Executive Order No. 4467, signed by President Calvin Coolidge on 29 July 1926 placing JA under the control of the Department of Agriculture as a breeding sanctuary for seabirds.

Many changes occurred in the Pacific prior to, and as a result of World War II. Johnston Atoll came under Department of the Navy control when President F. D. Roosevelt signed Executive Order No. 6935 on 29 December 1934. . At this time, the atoll still had only the two original islands, Johnston and Sand Island. Sand Island remained a bird sanctuary under the Department of Agriculture. However, increased military activity in the years from 1939 to 1941 included building of facilities on both islands with a small shallow channel dredged between them. Sand Island became the main occupied island while Johnston Island was expanded by dredging the lagoon coral in order to make a Naval Air Station. The increase in size was necessary to accommodate a longer runway and additional aircraft parking space. The station became operational in August 1941 and Executive Order No. 8682 (Febuary 1941) designated the area out to 3 miles from the atoll as a Naval Airspace Reservation and a Naval Defense Sea Area. Also during this time, Pan American Airways used the air-field for trans-Pacific flights between Hawaii and Asia.

During the war, the atoll was a vital strategic site as a refueling station for planes and submarines patrolling the pacific in defense of the Hawaiian Islands. The US forces on JA came under cannon fire by Japanese naval vessels on 15, 21 and 22 December 1941. Facilities were damaged but there were no casualties.

After WWII, JA continued as an air station and was eventually transferred from the Navy to the US Air Force in July 1948. Dredging the lagoon slowly expanded the size of the islands while also creating deeper and larger shipping channels. The islands were heavily used during the Korean War. A US Coast Guard Loran station was built on Sand Island in 1960. It functioned as the master Loran station for the central pacific until December 1992 when the Global Positioning Satellite navigational system became fully operational and rendered Loran obsolete. The former Loran station became our marine biological field station in 1993 and was used until base closure activities began in 2001.

During the Nuclear Test Era of the 1960s, Johnston Atoll became a prime test site and staging area because it was ideally located only several hundreds of miles from Hawaii but still isolated in all directions from other inhabited islands. In 1962, the US commenced a series of atomic tests. The USAF released operational control of the islands to the Atomic Energy Commission (AEC) and their military counterpart, Joint Task Force Eight (JTF-8). The atomic test series code-named DOMINIC included four high altitude and five low altitude successful tests followed by a few THOR missile misfires. Radioactive contamination of the islands and the lagoon from fallout resulted. The Defense Special Weapons Agency (formerly the Defense Nuclear Agency) is responsible for continuing radiological surveys and cleanup.

A second series of scientific surveys of Johnston Atoll were conducted by the University of Hawaii during 1963 - 64 (Brock eta. 1965, 1966). One focus of these studies was to evaluate the ecological impacts of lagoon dredging. The phenomenon called ciguatera was one of the main concerns. Ciguatera is a marine toxin that originates in a single-cell plant, a benthic dinoflagellate. This dinoflagellate lives on the surface of other algae, which are eaten by herbivorous fishes which can then accumulate the toxin. Depending on the level of toxicity, humans eating ciguateric fishes can become mildly ill to the extreme of death. One theory for ciguatoxicity outbreaks is that environmental disturbances such as dredging and severe storms can foster blooms of the ciguateric dinoflagellate.

A comprehensive study of the seabirds with extensive notes on other aspects of the atoll's natural history was published in 1976 in the Smithsonian Institution's publication, *Atoll Research Bulletin* (Amerson and Shelton 1976). This publication also included a detailed history and summary of research findings up until that time.

In 1971, the US Army began using 41 acres (0.16 km2) of Johnston Island for chemical weapons storage. These weapons included nerve gas and mustard agent contained in rockets, artillery shells, bombs and ton containers. In 1972, the USAF brought about 25,000 55-gallon drums (25,000 208 l drums) of the chemical, Herbicide Orange (HO) that originated from Vietnam and was stored on Okinawa. This stock of HO was incinerated at sea in 1977 aboard the Dutch ship, *Vulcanus*.

During the years 1970 to 1985, the population of the island averaged about 300 people. This included military and civilians. The building of the Johnston Atoll Chemical Agent Disposal System (JACADS) plant began in 1986 and with its completion, the island population grew to its present level of about 1200 personnel. The USFWS stationed a full-time refuge manager on the island with Army funding starting in 1990 to help manage the natural resources. The increased number of people on the island also resulted in an increased exploitation of fish and marine invertebrates. Some species such as whitetip sharks and certain marine mollusks (especially species with attractive shells) were severely over-collected. Shark fishing and shell collecting activity has since been controlled.

Today, the main base of operations, including a complicated administrative organization involving officers and representatives from several armed services plus the US Fish and Wildlife Service is located on Johnston Island . The US Air Force at Hickam AFB, Oahu, HI, owns the land. Island management and overall military control is under a USAF Colonel, who supervises the islands vast accommodations and base operations supporting a workforce of approximately 1,200 people . The typical utility and maintenance facilities necessary to support everyone on the atoll are the main tasks of the base contractor. All freshwater is made by a desalination plant and power is generated from burning JP5 jet fuel. Today there are about 300 buildings and facilities in operation on the nearly 1.4 million square feet of space which comprises Johnston Island.

The island's military force of about 200 Military Police (MP) is under the command of an Army Colonel whose chain of command is the US Army Pacific, Ft. Shafter, HI. The Army MP mission is to secure the chemical stockpile from being tampered with or misused by unauthorized personnel. While chemical weapons were stored on the island, these soldiers had standing orders to use deadly force in its defense. The Army also had responsibility to store and transport the chemical weapons from the storage bunkers and to convey these weapons of mass destruction into the JACADS facility for incineration back into elemental form and scrap. Since the completion of munitions destruction activities, this organization is being inactivated.

The Army's Program Manager for Chemical Weapons Disposal Agency oversees the overall incineration business and operations. The plant was built and is currently operated by Washington Demil Company. There has been a long history of other contractors involved in all aspects of building and operating the facility. Gary McCloskey has been the Army's field office program manager on Johnston Atoll since the beginning and he has directly supervised the nations first environmentally correct chemical weapon incineration plant from conception to completion. JACADS began destroying chemical weapons in 1990 and completed its mission in 2001. The JACADS facility successfully destroyed over four million pounds of nerve and mustard agents contained in various weapons ranging from rockets to mines. By using sophisticated pollution abatement systems and other engineering controls, the Army has successful destroyed these weapons of mass destruction without any release of chemicals into the reef environment. For more information about JACADS, see the Army's website http://www.pmcd.apgea.army.mil.

Biogeography

The fish fauna of Johnston Atoll is comprised of about 301 species (Randall et al. 1985, Kosaki et al. 1991). In contrast to the 301 species found on JA, about 536 species are found in the Hawaiian Islands, about 844 species In the Mariana Islands and about 1,357 species in the Palau and Yap Island groups (Randall 1992). The lower number of species on JA is probably the simple result of there being fewer types of habitats compared to Hawaii or other archipelagoes. The close connection between Johnston Atoll and Hawaii is evident in the similar biogeography of shore fishes and reef corals. The majority of Johnston Atoll species are Hawaiian although some Line Islands species occur at Johnston Atoll but not in Hawaii. When one examines the distribution of Hawaiian endemic fish species and subspecies with their closest relatives from the Line Island (there are eleven ibling species pairs) we find that the Hawaiian form has colonized Johnston Atoll not the Line Islands form, with only one exception (Randall et al. 1985). The pattern of Johnston's fauna being dominated by Hawaiian species is further corroborated by finding only 17 non-Hawaiian fish species compared to 51 endemic Hawaiian species occur at Johnston Atoll (Randall et al. 1985, Kosaki et al. 1991). Coral species at Johnston Atoll are also predominantly Hawaiian with 29 species occurring at Johnston compared to about 40 in Hawaii (Maragos and Jokiel 1986).

Two pieces of additional evidence suggests a path of current drift and therefore larval transport between the Hawaiian Islands and Johnston Atoll and vice versa. Flow from the main islands to the Atoll was suggested by finding kukui nuts on Johnston Atoll beaches. The only known sources of this nut in the North Pacific are the main Hawaiian Islands and it was clear that these nuts had been adrift at sea before being washed ashore at Johnston. Kukui nuts were found during each of three field trips in 1983-84 (Lobel 1985) and have been seen in subsequent years as well. Other biogeographical data suggest current flow from Johnston Atoll to the northwest Hawaiian Islands (Grigg 1981; Grigg et al. 1981). Populations of a *Acropora* coral living at French Frigate Shoals and nearby reefs. do not appear to be sexually reproductive, and according to Grigg (1981), probably derive from larvae spawned at Johnston Atoll (see also Maragos and Jokiel 1986). The key point is that for such close faunal affinities to occur between Johnston Atoll and Hawaii, currents must exist which transport pelagic larvae from one location to the other. Prevailing currents from other directions could have connected Johnston Atoll to several other island groups, each with their own distinct fauna, but this has not apparently happened. Johnston Atoll is one of the most isolated Atolls in the world; however, it is situated within the domain of an Archipelago-sized ocean current gyre feature (illustrated in Wyrtki 1975 during March-April). These lines of evidence support the notion of an oceanographic link between Johnston Atoll and the Hawaiian Islands.

The phenomenon that 25% of Hawaiian shore fishes (total approximately 536 species) are taxonomically recognized as endemic species (Randall 1992, 1995) emphasizes their restricted biogeographic distributions. Given that many of these species have planktonic larval phases lasting up to three months (e.g., Ralston 1981, Lobel 1989, 1997), it is enigmatic how these fishes remain restricted to Hawaiian waters. One hypothesis argues that it is the synchrony of fish spawning patterns in Hawaiian waters to ocean eddy current patterns which entrains and retains pelagic larvae near home islands (Lobel 1978, 1989, 1997; Lobel and Robinson 1983, 1986). Limited evidence also suggests that JA may also have some degree of local retention of pelagic larvae and that the fish populations on the atoll are mostly derived from the breeding stock on the atoll rather than from larvae drifting in from other island groups (Lobel 1997). Several Johnston Atoll fishes exhibit color patterns suggesting they are sub-species and therefore distinct from their sister species in Hawaii. These JA fish populations that appear distinctly colored from their nearest relatives in Hawaii include a surgeon fish *(Ctenochaetus strigosus)*, a parrotfish *(Scarus perspicillatus)*, a wrasse *(Labroides phthirophagus)* and two butterflyfishes *(Chaetodon multicinctus and C. tinkeri)* (Randall 1955, Randall et al. 1985, Kosaki et al. 1991). One fish at JA, the pygmy angelfish, *Centropyge nahackyi*, is recognized as endemic to the atoll (Kosaki 1989). Additionally, another population unique to Johnston Atoll is a fish that appears to be a hybrid between two sister species; the Hawaiian endemic wrasse *Thalassoma dupperry* and its nearest relative in the Line Islands and West Pacific, *T. lutescens* .

Faunal statistics (numbers of species and genera) reported in the species accounts were extracted from data reported in Kosaki et al. 1991. Fisheries statistics and population estimates are from Irons et al. 1990

In order to generally describe the geographic range over which a fish species is found, a list of standard definitions is given below.

Circumtropical—Found worldwide in all tropical marine waters including the Pacific, Indian Ocean and tropical Atlantic, including the Caribbean.

Indo-Pacific—Wide ranging in the warm waters of the Indian Ocean and the central and western Pacific.

Indo-Pacific and tropical eastern Pacific—Ranges from the Indian Ocean throughout the Pacific to the tropical west coast of the Americas.

Hawaii and Johnston Atoll endemic—Occurs only in Hawaii and Johnston Atoll.

Western Pacific—The west Pacific boundary is in the area of the Andamans and Cocos Keeling Islands in the west, to the margin of the Pacific plate in the east, north to southern Japan and south to western and eastern Australia.

Central Pacific—Includes the island groups; Line Islands, Phoenix Islands to the Marshall and Caroline Islands, all north of the equator.

Anti-equatorial—Hawaii to southern Japan in the north and Easter Island to New Caledonia in the south. Absent along the equator.

Oceania—All of the tropical islands in the Pacific from Hawaii to the West Pacific in the northern hemisphere and from Tahiti to the Coral Sea in the southern hemisphere.

Tropical eastern south Atlantic—The tropical east coast of southern Africa.

Acknowledgements

My research group has been evaluating the impact of human recreational activities and the potential adverse ecological impacts resulting from past military and current JACADS activities since 1983. This project has been supported and funded primarily by the US Army Program for Chemical Demilitarization as part of the JACADS environmental monitoring project and was supplemented by the US Air Force Pacific Command, the Defense Threat Reduction Agency (formerly the Defense Nuclear Agency), the US Coast Guard, and the Legacy Resource Management Program. This work was supported by the U.S. Air Force Pacific Command as part of the "Former Herbicide Orange Storage Site" Environmental Risk study and by the U.S. Army Chemical Material Destruction Agency as part of the JACADS Marine Environmental Monitoring Program. Funding administration was provided by the Army Research Office (Grant DAAG55-98-1-0304), the Office of Naval Research (Grants N00014-19-J1519, N00014-92-J-1969 and N00014-95-1-1324) and NOAA (DACA83-83-0051 and NA90-AA-D-SG535). This publication was supported by Legacy Resource Management Program (grants DACA87-00-H-0021 and DAMD17-93-J-3052 and the Army Research Office (Grant DAAG55-98-1-0304).

On Johnston Atoll, we have been fortunate to have the friendship and support of many people, particularly the staffs of the JACADS field office, the Defense Special Weapons Agency; the US Air Force and the U.S. Fish and Wildlife Service. We are indebted to the following individuals for their help in many ways over the past several years (in random order): Gary McCloskey, Mary Demmitt, Peggy O'Connell, Dave Shogren, Carey Akamine, Dan Fukuoka. Mark Ingoglia, Roger DiRosa, Jay Silberman, Lindsey and Beverly Hayes, Donna O'Daniel, J. Bruce Owens, Robert Campbell, Jeff Klein, George Ringle, "Stu" Studebaker, Roger Rios, "Smitty" and Bob Kunukau. I am extremely appreciative of the many people on the staffs of the Legacy Program, PMCD, USAF Pacific and Army Research Office for their longtime technical and logistic support and for taking an interest in this research.

I am also very grateful to the faculty of Zoology and Oceanography Departments at the University of Hawaii for their constant help and encourgement over the years since I began as an undergraduate there in 1971. The UH connection got me involved with Johnston Atoll in the first place. I was first introduced to Johnston Atoll by an invitation from James Maragos and E. H. (Deetsie) Chave to participate in the research expeditions of 1983 using the NOAA research submersible, MAKALI'I, to map and document the atoll's marine resources. This expedition also included John (Jack) Randall. Deetsie, Jack and Jim have been helping me ever since with the Johnston project as well as other projects and I am most grateful to them for the many years of mentorship and friendship.

I also extend many thanks to the entire team of students and colleagues who have worked with me on the atoll over the past 18 years, including Anne Cohen, Randy Olson, Skip Carter, David Mann, Gorka Sancho, Steven Oliver, Erick Horgan, Kent Merrill, Gorka Sancho, Alistair Economakis, Jason Philibotte, Gaby Tomasky, Les Kaufman, Becky Maxson, Justin Neviakas, Greg Skomal, Spence Smith, Elise Watson, and many others. Conducting Research on the lonely atoll became much more fun and productive when Lisa M. Kerr joined the project in 1994 and the best result was when she became my wife. If not for her help in every facet of the project, I doubt that I could have completed as much science as we have done. This project could not have been completed without Russ Carpenter and his patient supervision and ultimate publication of this book.

Johnston Atoll as viewed from the east looking west. This view has East Island in foreground on lower left and North island in foreground on lower right. Sand Island is in the center and Johnston Island is the largest in the background. The dredged areas are visible next to North Island and leading to East Island. The emergent barrier reef is on the north margin of the atoll. The south atoll margin is submerged and is without a barrier reef formation.

Johnston Island was originally a mere 46 acres (0.18 km^2). Its expansion by dredging began in 1939 and by 1942 it was 211 acres (0.85 km^2). The second expansion took place in the 1960s and the island grew to about 624 acres (2.58 km^2). This view is looking south. The emergent barrier reef is in the foreground and the typical "spur and groove" geology is visible. The inner lagoon is packed with extensive shallow reefs. The dark water near the island is the channel. Dredging and crushing live coral, during an era when environmental concerns and regulations were minimal, was the method use to build the islands. Johnston Island is about two miles long and a half-mile wide. An airplane runway runs most of the length. The ship tied to the dock is one that periodically brings supplies and is about 600 feet (180 m) length. The JACADS facility is situated on the small peninsula on the opposite side in the middle. The other peninsula on the east end (left) is used for hazardous waste storage before it is removed from the atoll by cargo ship.

The herbicide used to defoliate trees in Vietnam, Agent Orange, was stored on several acres located on the northwest corner of Johnston Island. Approximately 250,000 lbs (113,400 kg) of Agent Orange leaked and contaminated the adjacent reef area with dioxins, but the effects are apparently very localized.

JACADS facility. Johnston Atoll Chemical Agent Disposal System is a facility of the US Army to destroy chemical weapons most of which are left over from WWII. The facility is under the direction of the Army's Program Manager for Chemical Demilitarization.

JACADS is located on a peninsula in the middle of Johnston Island. Construction of the plant began in 1985 and chemical demilitarization (incineration) operations began in 1990.

This scene shows the marina area's main wharf. The buildings near the fuel silos include the desalination and power plants. The motor pool is on the right and three housing buildings are in the lower left of the photograph.

Downtown Johnston island. The lowrise building across from the parking lot is the "Coral Reef Café", the main dining facility. The other buildings include apartments. The airfield is in the background.

North Island, shown in the foreground, is another artificial island created by dredge and fill. Sand Island, shown in the background, was doubled in size by dredging. The east or original side of the island is the main site for seabird nesting within the atoll. The Boston University Marine Laboratory, located on the causeway was established for marine ecological monitoring of JACADS and the other military actions past and present in the atoll.

East Island was created entirely by dredging the adjacent coral reef. Overall, the dredging operations in the 1960's destroyed over 1,100 acres (4.5 km^2) of reef, in addition another 7,000 acres (28.3 km^2) of reef was adversely impacted by silt and murky water that resulted.

Sand Island was once the site of a US Coast Guard LORAN Station. It became a marine lab in 1993 after GPS navigation made LORAN obsolete. It was the hub of marine research activity for Johnston Atoll until 2001 when as part of Base closure process; the marine lab was moved to Johnston Island.

This view of Sand island shows the eastern half of the island and the causeway. The building is the marine lab. The area in the foreground, along the southern shore of the east half of Sand Island is the location where female grey reef sharks aggregate annually (see photo 39).

While Johnston Atoll is a military site on the surface, it contains lush coral gardens underwater. The reef ecosystem, in general, is healthy and in very good condition.

The tugboat wreck is a favorite dive site for introductory lagoon diving. There are a few sites in the atoll with old wrecks and military debris. All harmful materials (e.g. batteries, transformers) have been removed.

Shallow reef ridge near Sand Island..

Deep reef formations in the center of the lagoon.

The top photo shows a WWII half-ton truck that probably fell off the seaplane ramp (in background) and was never recovered. The lower photo shows rusted 55 gallon drums among other debris that is scattered on the bottom near the old wharf on Sand Island. This area was found to have low levels of PCBs in the sediments and the impact from this contamination is under study by Lisa Lobel.

Corals

Stony corals are the dominant substrate in the lagoon and outer reef slopes. Overall coral diversity is low as compared to other tropical atolls. However, the coral cover in the lagoon is "wall to wall" and dense except where patches of sand occur. About 42 species have been recorded on Johnston Atoll. The species illustrated here are the most abundant.

Plate coral, Montipora capitata, with Moorish Idol, Zanclus cornutus.

Mushroom coral, Fungia scularia is abundant on the reef crest.

Yellow-lobed coral, Porites lobata

Bush coral, Acropora humilus

Cauliflower coral, Pocillopora meandrina

Finger coral, Acropora cerealis

The shallow lagoon is dominated by coral interspersed with sand patches. The water is generally clear most of the year. Sea grass, mangroves and other habitat types are not found at Johnston Atoll.

A typical Johnston Atoll patch reef.

Fire coral, *Millipora tenera.*

Soft coral, *Sinularia abrubta.* Soft corals are found only in areas of strong current flow on the outer reef slopes and in some of the channels through the reef.

Red coral, *Distichopora sp.* Semi-precious red coral is quite abundant on the outer reef slopes. It grows in small colonies rarely larger than a small plate in size. All corals within the atoll are protected and collecting is prohibited.

A massive coral spawn occurs annually in late May to early June. Locally known as "red tide", this is not related to the true red tide caused by dinoflagellates (microalgae) blooms. The red material is actually tiny coral larvae that are produced in huge quantities over a few day period. The odor of the coral spawn can be so strong and pungent that it once set off the JACADS mustard agent detectors.

The dominant coral in the lagoon, *Acropora cytherea* is a fast growing species that has a growth form like a tabletop. This is the most abundant coral on Johnston Atoll, but has rarely been observed around the Northwestern and main Hawaiian Islands.

Macroalgae

A total of 93 different species of marine algae have been found on JA. However, most of these are small and cryptic and are cropped to a low stubble by grazing herbivorous fishes. Only a few types of algae, which are generally unpalatable to herbivores, are seen growing in noticeable mass.

CYANOBACTERIA, BLUE-GREEN ALGA, possibly *Schizothrix* sp (Oscillatoriaceae). The two blue-green algae are found mainly in the deep lagoon near Johnston Island. Both overgrow and smother living coral. Herbivorous fishes do not apparently eat these particular algae. There are several very similar species, the identity of the JA species has not yet been determined. It grows as a thick mucous mass. Two common species found throughout the tropical Pacific Ocean are *S. calicola* and *S. mexicana*. The two species differ in the diameter of their filaments (S. Mexican is thicker).

CYANOBACTERIA, BLUE-GREEN ALGA, possibly *Microcoleus lyngbyaceus*). This alga also grows as a thick mat on the reef. Its mats are usually larger and less mucousy than *Schizothrix*. It can also be very red to brownish in color. This alga is found throughout the world's tropical seas. The thick mats can trap bubbles of oxygen, which can float mats away.

CALCAREOUS GREEN ALGA, *Halimeda opuntia* (Hallmedaceae) Circumtropical, grows to approximately 4 inches (10 cm) height. This alga is also conspicuous on reefs. It is easily identified by its segmentation and obvious calcification. The specific shape of the segments helps distinguish species. Calcium carbonate from *Halimeda* spp is a significant source of sediments on tropical reefs worldwide.

GREEN CORKSCREW ALGA, *Caulerpa serrulata* (Caulerpaceae). This species is found in all tropical oceans including, Pacific, Indian Ocean and Caribbean reefs. It grows approximately 1 to 3 inches (2 to 8 cm) high. This species is the dominant macroalga in the Johnston Atoll lagoon, especially in the winter season. In the area along the south shore where green sea turtles are most frequently observed, this was the only macroalga found, but it was abundant, forming large mats. This species is also used as a human food and as a traditional medicine due to its antifungal properties and ability to lower blood pressure (Trono 1997).

Spiny sea cucumber, *Stichopus chloronotus*

Echinoderms:

Sea Cucumbers, Sea Urchins & Sea Stars

There are about 37 species belonging to 13 families of Echinodermata known on JA. These following are the species most often encountered by divers.

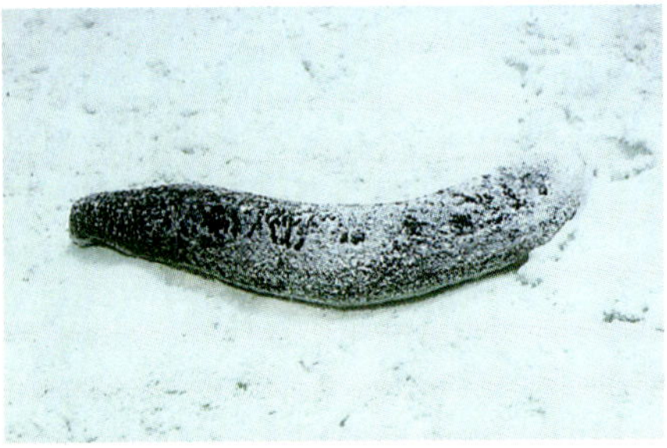

Black sea cumber, *Holothuria atra*

Sea cucumber, *Holothuria edulis*

Rock-boring urchin, *Echinometra mathaei*

Pincushion urchin, *Tripmeustes gratilla*

Crown-of-thorns seastar, *Acanthaster planci*

Freckled seastar, *Linckia multifora*

Seabirds

Terns and noddies nest in huge numbers on Sand Is. The Atoll is habitat for 15 species of seabirds. Overall, about 56 bird species have been observed at least once on the atoll including mallard ducks, owls and a falcon. Unlike birds found in the Northwest Hawaiian Islands, none of the birds on Johnston Atoll are endemic or threatened species.

Sotty tern, *Sterna fuscata*, showing the nesting population.

Brown Noddy, *Anous stolidus*, several thousand nest on the outer islands. They lay a single egg which is cared for by both parents.

Red-tailed tropic bird, *Phaethon rubricauda*. Are similar to the white-tailed tropic bird and differ, of course by the color of the tail feathers. Both species nest on the atoll.

Brown booby, *Sula leucogaster*..

This species lays two eggs but only one chick survives due to sibling homicide. Once hatched the chicks grow rapidly and attain adult size in a few months. The white bird is the chick and already is adult size while still being cared for by the parents. These birds feed on squid, flying fish and larval reef fishes.

Great frigate bird, *Fregata minor palmerstoni.*

The males are easily distinguished by the red throat pouch. These birds can reach a seven foot wingspan and are the largest species on the atoll. Great frigate birds are well known as pirates who steal food from other seabirds.

These birds can be seen nesting on Sand and North Islands.

Octopus (Cephalopoda)

DAY OCTOPUS, *Octopus cyanea,* (Octopodidae). Distribution: Indo-Pacific. Max 4 to 5 lbs (9-11 kg), length 3 feet. (95 cm).

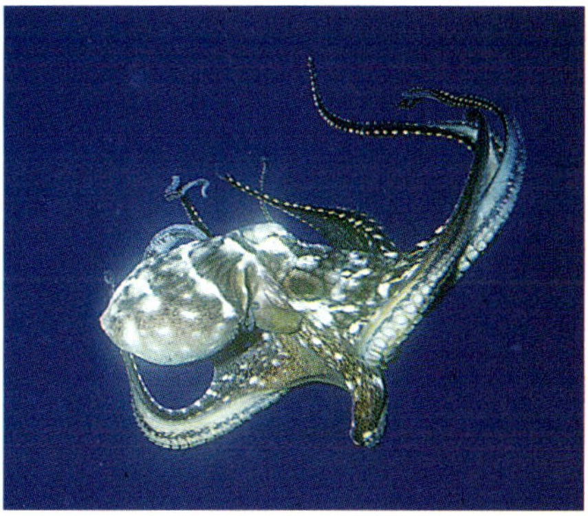

Octopus posed in midwater, about to make its escape.

Swimming, using jet propulsion.

Camouflage includes rapid color change with modification of its skin surface texture.

Ejecting ink cloud as last ditch effort to escape a threat.

Seashells (Gastropoda)

TIGER COWRY, *Cypraea tigris* (Cypraeidae). Distribution: Indo-West Pacific, including Hawaii. Size: max 6 inches (15 cm).

The tiger cowry is found throughout the Pacific Ocean, but reaches its largest sizes in the Hawaiian Islands (up to six inches/15 cm). It is an herbivore and mainly feeds at night on the reef and hides during the day. The shell can be highly variable in color pattern and may be white or golden brown with scattered dark brown or black spots. Ancient Hawaiians used the tiger cowry as a lure to catch octopus.

TRITON'S TRUMPET, *Charonia tritonis* (Ranellidae). Distribution: circumtropical. Max length 18 inches (46 cm).

This is largest seashell found on Johnston Atoll. It was heavily collected by recreational divers until DoD implemented coral reef protection policies that prohibit souvenir shell collecting.

The tritons are found mainly in reef areas. They are best known as predators on the crown-of-thorns sea star but they also eat other sea stars and sea cucumbers. Tiger sharks have been found with tritons in their stomach.

Johnston Atoll has an abundant diversity of marine mollusks with many of the same species are are found in the Hawaiian Islands. A few of the common species are shown here.

AUGERS (left to right)

WHITE SPOTTED AUGER, *Terebra guttata*

MARLIN SPIKE AUGER, *Terebra maculata*

AGATE AUGER, *Terebra archates*

TUN SHELLS: APPLE TUN, *Malea pomum* (left); PARTRIDGE TUN, *Tonna perdix;* (center); and the VIBREX HELMET SHELL, *Casmaria erinaceus kalosmodix*..

Sea Slugs or Nudibranchs (Opisthobranchia)

These marine gastropods lack an hard external shell. There are several species seen at JA, the two illustrated here are very common and co-occur in the area offshore the former Herbicide Orange storage area on the northwest corner of the island.

Neon nudibranch, *Chelidonura hirundinina*. Length 1 inch (2.5 cm). Distribution: circumtropical. This nudibranch is a predator on very small acoel flatworms. This type of flatworm is often found in association with corals and sea anemones and some have symbiotic algae in their tissues. This nudibranch swallows its prey whole.

New neon nudibranch, *Chelidonura sp.* Length 1 inch (2.5 cm). Distribution: Johnston Atoll and some other Pacific islands. This is a scientifically undescribed species which is common at JA. This new species is under study by T. Gosliner (California Academy of Sciences, San Francisco).

Hawaiian Monk Seal

HAWAIIAN MONK SEAL, *Monachus schauinslandi.* Distribution: Found mainly in the Northwest Hawaiian islands. On a few occasions, Monk seals have appeared on the shores of Johnston Atoll but historically have never been known to establish long-lasting colonies. Lost individuals, apparently wandering far off their normal range have ended up at JA. The National Marine Fisheries Service have also transplanted excessively aggressive males from the breeding colonies located in the northwest Hawaiian Islands to JA because they were rogues and responsible for killing pups. None of the Monk seals moved to JA have remained for more than a few months. The female shown above appeared on her own in 1998 with a massive tiger shark bite wound on her belly. She was seen about the atoll for several months before disappearing. She had been tagged by the NMFS on French Frigate Shoals in the Northwest Hawaiian Islands.

Hawaiian Monk Seal, *Monachus schauinslandi*

Green Sea Turtle

GREEN SEA TURTLE, *Chelonia mydas*. Distribution: Found in all tropical seas. Green sea turtles arrive at Johnston Atoll throughout the year to feed but they appear to return to the northwest Hawaiian Islands, especially French Frigate Shoals to breed. One nest was found on JA but the eggs did not hatch. The green sea turtle is seen most frequently along the south side of Johnston Island where it is shallow and macroalgae grow abundantly.

Green Sea Turtles are sometimes found under ledges where they have hidden to rest.

Sharks (Carcharhinidae)

Big sharks are rarely seen in or around Johnston Atoll. A thresher shark (*Alopias sp.*, Alopiidae) was seen once at 600 feet (183 m) depth by observers in the University of Hawaii Makalii research submarine in 1983. We have seen a few Galapagos sharks, *Carcharhinus galapagensis* (Carcharhinidae) on the deep reef ledge. A scuba diver once observed a hammerhead shark, (possibly, *Sphyrna lewini*) on the outside reef slope.

The one large shark that is seen most frequently, especially in the deep main channel into the lagoon, is the tiger shark (*Galeocerdo cuvier,* Charcharhinidae). Tiger sharks are most likely to be seen in the deep channels during early evening and nighttime. The best location to observe tiger and grey reef sharks on JA is at the west wharf of Johnston Island when the leftover food from the "Coral Reef Café" is disposed by ocean dumping. Tiger sharks are also known to follow ships that discard food refuse. On JA, tiger and Galapagos sharks follow the fishing boats and it is often a race to bring the fish in before the sharks steal the catch. Their natural diet includes other sharks, rays, turtles, porpoises, dolphins, sea birds, lobsters, tritons and most other seafood as the opportunity presents itself. Tiger sharks grow to more than 15 feet (4.6 m) in length. Females mature at about 10 feet (3 m) and have as many as 80 pups (1.5 to 3 feet or 0.4 to 0.9 m in length) in a litter. The tiger shark is distributed circumglobally in tropical and warm water seas. Tiger sharks have been responsible for many of the fatal attacks on swimmers and surfers in Hawaiian waters.

GREY REEF SHARK, *Carcharhinus amblyrhynchos* (Carcharhinidae). Distribution: Found throughout the tropical Indo-Pacific Ocean, from the Red Sea across the Indian and South Pacific Oceans, including Australia's Great Barrier Reef and out to Lord Howe Island, throughout the Central Pacific Ocean from Tahiti to Hawaii, eastward to Easter Island. Max. size 7 feet (215 cm). On Johnston Atoll, this species is most likely encountered along the outside reef and in channels through the barrier reef connecting the lagoon to deep water. Numerous grey reef sharks were once commonly seen aggregated in feeding frenzies off the west wharf of Johnston Island whenever the dining hall refuse was dumped into the lagoon. A male can be recognized by the two claspers protruding from between the pelvic fin and extending to the anal fin. The male inseminates a female internally using a clasper.

GREY REEF SHARK, *Carcharhinus amblyrhynchos.* A female can be recognized by her lack of claspers. The female has about a 12 month gestation period and typically produces 4 pups.

This species is well known for its complex behaviors including an aggressive display. The grey reef shark exhibits its displeasure with being crowded or chased by a diver (such as when trying to take a close photo!) by arching its back, lowering its pectoral fins and swimming in an erratic zigzagging pattern. The shark in the photo is displaying a mildly aggressive posture. Threat displays are easy to recognize if one remembers to pay attention to the shark's swimming behavior. Once a shark is agitated, the shark changes its swimming behavior from seemingly effortless gliding through the water into a distinct posture with a curved and twisted body and with pectoral fins extended downward. The contorted shark in threat display swims in an exaggerated swimming motion where it swings head and tail together on one side and then the other. The distinctive threat display may be followed by an attack which usually results in a single non-fatal bite. To my knowledge, no scuba diver has yet been attacked by a grey reef shark on Johnston Atoll. In general. it is best to remain calm and to face, but not swim toward a grey reef or any other shark. Spear fishing of course, attracts shark's special curiosity and up close personal inquiry that can lead to a dangerous encounter.

Grey reef shark, *Carcharhinus amblyrhynchos.* Female grey reef shark aggregation at Sand Island. A small group of about 24 individuals.

Grey reef shark, *Carcharhinus amblyrhynchos.* Sand Island, aggregating females underwater at high tide. Many females appear to have big bellies and are probably pregnant.

A female grey reef recently mated shows the scars from a male's love bites.

The grey reef shark is one of the atoll's most interesting and watchable creatures. When diving on the outer slopes it is not unusual to have anywhere from three to thirty greys swarming around, their graceful ghost-like movements in and out of visual range makes it difficult to keep one's eyes on anything else. Once a year, starting in late Febuary and continuing into early May, female greys migrate into the lagoon and into a shallow site off Sand Island. The females arrive daily about mid-morning and depart around sunset every day during the aggregation period. They school daily, mostly during daylight hours, but reappear sometimes at nighttime. The female grey reef sharks appear in numbers ranging up to 150 individuals swarming in a shallow area of only a few acres. Progressing into May, the numbers dwindle until only 12 to 20 appear each day and then numbers continue to dwindle until none appear. Only females have been observed in these aggregations and of the few that have been examined, all have been pregnant with from two to five embryos in a single female.

The gestation period is about one year for this species and litter sizes range from one to six pups. The newborn pubs are born fully developed and are about 2 feet (0.6 m) in size. Grey reef sharks mature at seven years and probably live to about 25 years. A mature shark measures about 4 feet (1.2 m) in length but individuals may grow to 7 ft (2.1 m) or more.

The reasons for their aggregation are the subject of ongoing scientific studies examining the reproductive biology and migration patterns of this species. The natural history story behind the behavior of female grey reef shark aggregations is not well known. Aggregations of female grey reef sharks have been seen on other islands and atolls throughout the Pacific Ocean. In all cases, it is only females that swarm together in the shallows and a portion, but not all, of the female population are pregnant. Divers have been known to swim among the female sharks without being attacked and it has been speculated that females will not feed at this time. However, pieces of tuna thrown to the aggregated females result in vigorous feeding frenzies.

Scientific study of behavior of the female shark aggregations off Sand Island is being conducted with research support from the US Army and the Legacy Resource Management Program of the Department of Defense. As part of this study, we measure a variety of oceanographic and meteorological variables at this site and elsewhere in the lagoon. We also track the movements of individual sharks by attaching radio transmitter tags. One aspect of this study is to determine if the shallow water area preferred by the female sharks differs in temperature, salinity etc., or in other factors from other areas in the lagoons. It is possible that the sun warmed shallow water may somehow help the metabolism of pregnant sharks. It is also possible that the aggregation is a refuge for females which have already mated or who are not ready to breed. By remaining in the aggregation, they avoid the amorous attention of the males, which are on the outer reef slope. Shark mating can be quite vigorous and it always results in the female acquiring a series of wounds from the male's love bites.

While most of what is known about shark behavior consists of observations of individuals swimming across a diver's field of view, grey reef sharks display several behaviors that suggest some type of social behavior. I have seen solitary males and males associated with other males and females, in addition to groups of three to 12 females roving together. One to three large females have been seen as if sheparding a flock of small ones. Several grey reef sharks have been observed hunting together as a pack by herding schooling fish up against the reef before attacking. On JA during January, large numbers of flying fish enter the lagoon in the west channel. Sharks have been repeatedly observed hunting in packs and chasing these flying fish toward shore. Grey reef sharks typically feed on reef fishes, flying fishes, squid, octopus, lobsters and crabs.

REEF WHITETIP SHARK, *Triaenodon obesus*. Distribution: Indo-Pacific, including the Red Sea, to the tropical eastern Pacific. The whitetip shark attains a size of little over 6 feet (1.8 m), has litters of one to five pups and grows at a rate of approximately one to two inches (0.5 to 1 cm) per year (Randall 1977).

The whitetip shark *(Triaenodon obesus)* is one of the top predators in the lagoon and it is especially adept at capturing prey among the dense colonies of tabletop corals *(Acropora cytherea)*. The whitetip shark population at JA was severely impacted by recreational fishing until marine biologists brought this issue to the attention of the US Fish and Wildlife Service and a ban on shark fishing was issued in 1992. Whitetips are a relatively docile shark and rarely threaten divers unless harassed or baited.

The whitetip shark is a reef habitat specialist and it is most frequently seen in the lagoon but can also found on the outside reefs. The whitetip can rest on the bottom and is frequently seen "sleeping" in caves or underneath huge foliose corals. It is more active during twilight and nighttime and typically hunts within a home area. This shark will feed on almost any easy prey including reef fishes (especially parrotfish and surgeonfish) and crustaceans. Tiger sharks and even giant groupers have been known to eat whitetip sharks.

Manta Rays (Mobulidae)

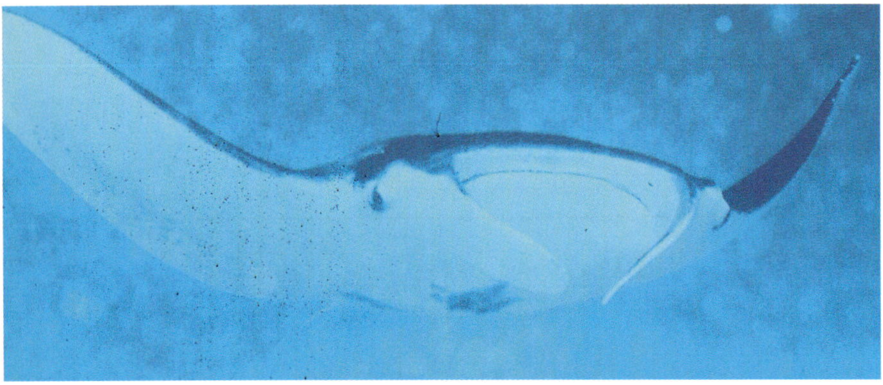

MANTA RAY, *Manta birostris*. Distribution: Circumtropical. Mantas are the largest of all rays and attain sizes of up to 21 feet (6.4 m) wide and weigh up to 3,000 pounds (1,360 kg).

Manta rays typically appear in the Johnston Atoll lagoon in May and remain off and on throughout the summer months. Their arrival is coincident with a change in weather and ocean current patterns that result in generally murkier lagoon waters caused by an abundance of plankton. The plankton includes the eggs and larvae of reef fishes and the mass spawn of corals, which usually takes place in the end of May. Very little is known about manta ray behavior and ecology. Two species are found within the Johnston lagoon, *M. brevirostris* is the most common. The other species, *M. japonica*, is rarely seen. Manta rays are highly variable in color pattern with each individual being distinctively mottled. Thus, it is easy to recognize individuals. Mantas are most commonly seen in the main channel off Hama Point on the southeast corner of Johnston Island. Manta rays feed on anything planktonic. They often feed just beneath the surface and observers sometime mistake the exposed wing tips for shark dorsal fins. Manta rays will breach, but their jumps rarely rise completely out of the water, as do eagle rays.

Manta Ray, *Manta birostris*. This individual was unusual in being mostly black on its ventral side.

Manta Ray, *Manta birostris.* The tail is relatively shorter than in *M. japonica.* Very little is known about manta ray biology. While different species can be identified taxonomically, it is unknown how sympatric species compare ecologically or behaviorally.

Japanese devilray, *Manta japonica*. Distribution: Indo-Pacific, and tropical eastern Pacific and possibly tropical west Africa. This species is most easily recognized by the much longer tail and more slender head profile. It is rarely seen at JA.

Japanese devilray, *Manta japonica* (Mobulidae)

Eagle Ray (Myliobathdae)

Spotted eagle ray, *Aetobatis narinari*. Distribution: Circumtropical. This species is found in every tropical ocean and sea around the world. Max. size 7.5 feet (230 cm).

Spotted eagle rays are seen frequently throughout the lagoon especially among the patch reefs in the deep channel between Johnston Island and North Island. Coloration can vary from black with white spots to solid black. It is usually solitary but sometimes Jacks (Carangidae) swim close to the dorsal side in between the wings of the eagle ray. The spotted eagle ray feeds on mollusks and crustaceans buried in the sand. Presumably, the Jacks nestle with the eagle ray in anticipation of being able to capture prey pushed out of cover by the eagle rays behavior. Eagle rays, like most other elasmobranch fishes, find hidden and buried prey by using specialized sensory pores found mostly on the head but also occurring all over their body. These sensory pores detect the electric field produced by living animals. The Spotted eagle rays can locate hard shell invertebrate prey buried beneath several inches of sand. The prey are extracted by the ray's digging which includes vigorous flapping of its `wings' to help flush away sand. Eagle rays can be seen probing the sand with their snouts. Evidence of a spotted eagle ray's foraging is the occurrence of deep (about 2 to 8 inches or 5 to 20 cm) depressions in the sand and the scattered hard parts of mollusks and crustaceans. Adults also feed on reef fishes. The eagle ray is most often seen swimming, unlike most other rays that mostly sit stationary on the bottom. The spotted eagle ray is a livebearer and has one to four pups in a litter. The gestation period is 12 months. During certain seasons at many locations worldwide, aggregations of up to dozens of eagle rays have been seen swimming together for days and across vast distances. Presumably, this is a social behavior and may be a prelude to mate selection and reproduction.

Eagle rays are prey for big sharks, most probably tiger sharks that hunt in the lagoon at night. Eagle rays protect themselves using one to five poisonous barbs that grow at the base of their tail.

Spotted eagle ray *Aetobatis narinari*

Predatory Reef Fishes

Snappers, Barracuda, Jacks, and Eels

This group of fishes includes members of several families, all of which are predators on macro-invertebrates and other fishes. In general, these fishes are seen cruising the reefs in search of an easy feeding opportunity. Eels are very abundant at JA and are usually seen with head protruding from a coral crevice.

Smalltooth snapper, *Aphareus furca* (Lutjanidae). Distribution: Indo-Pacific. Max. size 15 inches. In addition to the jacks, this reef snapper is one of the most common roving fish predators in the lagoon. They feed opportunistically on any small species but particularly have been seen hunting individual dwarf chromis (*Chromis acares*) from schools. There is a total of 9 snapper species (5 genera) at JA which comprise about 3% of the total fish fauna.

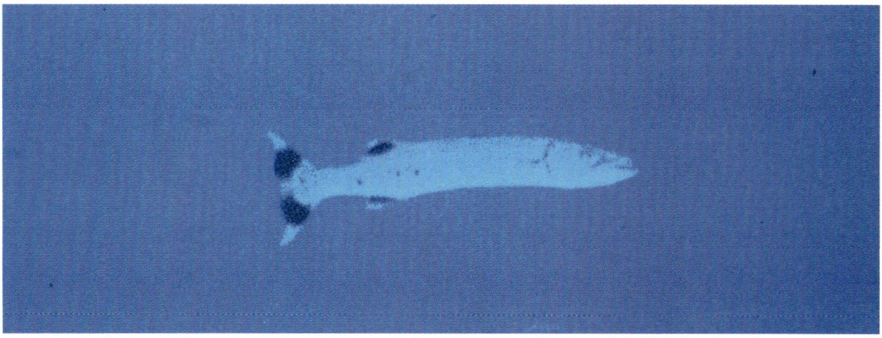

Great barracuda, *Sphyraena barracuda* (Sphyraenidae). Distribution: Circumtropical. Max. size 66 inches (168 cm). Two species of barracuda occur on the atoll and both are rarely seen. The other species on JA, *Sphyraena helleri*, is smaller and only seen on the outside reefs. *S. barracuda* is a predator on small reef fishes.

GIANT BARRACUDA, *Sphyraena genie (*Sphyraenidae). Distribution: Indo-Pacific. Max. size 40 inches (100 cm). This species is widespread in the equatorial Pacific and may stray to JA and possibly Hawaii (J. Randall pers. com.). It is slightly smaller that the great barracuda and is more frequently found in shallower water near the reef crest in the wave zone.

TELLOWFIN TUNA, *Thunnus albacares* (Scombridae). Distribution: worldwide in tropical and subtropical seas. Max. size 85 inches (210 cm). and approximately 200 pounds (485 kg.) Tuna are common offshore of JA and very rarely are seen by divers on the outside reef dropoff. These fish are predators on many types of small fishes and squids.

ONO OR WAHOO, *Acanthocybium solandri* (Scombridae). Distribution: worldwide in tropical seas. Max. size 85 inches (210 cm) and approximately 90lbs (198 kg.). The ono is a prized catch by recreational fishers on JA and the fish is abundant due to the low intensity fishing in local waters due to the protection by the military. This fish is rarely seen swimming along the deep reef dropoff along the northern atoll margin. At a distance it may be mistaken for a barracuda. The main difference to look for is the shape of the tail fin. The ono has a deeply forked and rigid tail. It is usually cruising at a good pace through the water. The barracuda has a broad and flexible tail, it usually is hovering or moving slow through an area.

Jacks (Carangidae)

There is a total of 11 jack species (7 genera) at JA, which comprise about 3.7% of the total fish fauna. All of the jacks are predators mainly feeding on small fishes and crustaceans including crabs and shrimps. Species not pictured include; *Carangoides orthogrammus*, *Decapterus macarellus*, *Scomberoides lysan*, *Selar crumenophthalmus*, and *Seriola dumerili*.

BLACK TREVALLY, *Caranx lugubris*. Distribution: Circumtropical. Max. size 36 inches (91 cm). This species is relatively common in the lagoon. One of the places this fish is commonly seen is under the sea-plane ramp on Sand Island.

GIANT TREVALLY, *Caranx ingnobilis*. Distribution: Indo-Pacific. Max. size 65 inches (165 cm). The very large jack in this photo was seen daily at the Port Control dock over several months. Once it appeared displaying scars and with the top of its tail bitten. It disappeared soon after.

Bluefin trevally, *Caranx melampygus*. Distribution: Indo-Pacific and tropical eastern Pacific. Max. size 39 inches (99 cm). This is probably the most abundant species of jack in the lagoon. Although it grows large, most individuals seen are small and are voracious predators on small benthic fishes. This species is one of the top ten fishery species and has an estimated population size of 26,350 individuals in the lagoon.

Barred trevally, *Carangoides ferdau*. Distribution: Indo-Pacific. Max. size 21 inches (53 cm). This species is usually seen on the outer reef in groups of a few to several fish.

Bigeye trevally, *Caranx sexfasciatus*. Distribution: Indo-Pacific and tropical eastern Pacific. Max. size 33 inches (84 cm). This species has been seen most often during the late afternoon in large schools that reside near a gap in the reef. Large schools of this and other jacks are rarely seen today in Hawaii or other Pacific Islands where fishing is common.

Rainbow runner, *Elagatis bipinnulata*. Distribution: Circumtropical. Max. size 48 inches (122 cm). This is an offshore species rarely seen by divers. When sighted it is usually on the outside reef along the dropoff.

Eels (Congridae, Muraenidae, & Ophichthidae)

Eels are the most speciose group of fishes on JA. There are 30 moray (Muraenidae) species (in 7 genera) at JA, which comprise about 10% of the total fish fauna. The next most common eel group is the snake eels (Ophichthidae) with 10 species (7 genera) comprising 3.3% of the total fish fauna. In addition, there is a worm eel (*Moringua ferrugatus*, Moringuidae), two conger eels (*Conger oligoporus* and *Conger cinereus*, Congridae see below) and one species of false moray (*Kaupichthys hyoproroides*, Chlopsidae). All eels are carnivorous and feed on a wide variety of small fishes and invertebrates.

SNAKE EEL, *Myrichthys magnificus* (Ophichthyidae). Distribution: Hawaii and Johnston Atoll endemic. Max. size 31 inches (79 cm). This species is very similar to *M. maculosus*, which is found elsewhere in the Indo-Pacific. This fish can always be seen at the fish cleaning station by the main wharf, especially when fish entrails are disposed in the water.

Several snake eel species have been collected at Johnston Atoll but only *Myrichthys magnificus* is commonly observed. Other species are rarely seen and have been collected by ichthyologists using special chemicals to coax them out of the reef. The species known from Johnston Atoll are: *Brachysomophis sauropsis*, *Leiuranus semicinctus*, *Muraenichythys cookei*, *M. gymnotus*, *M. schultzei*, *Myrichthys bleekeri*, *Phyllophichthus xenodontus*, *Schizmorhynchus labialis*, and *Schultzidia johnstonensis*. Snake eels are secretive and most species are either nocturnal or confine their activities to the hidden interior of reef formations.

Snake eels differ from moray eels in several notable features. Snake eels are usually round in cross section, have extended snouts with tubular nostrils, and the jaw is typically underslung. They do not typically keep their mouths open when their head protrudes from a crevice as do many moray eels. Snake eels have a stiff pointed tail, which they use to burrow backwards into sand or to prod their way into crevices. These fish have a keen sense of smell and feed on small fishes, crustaceans and worms that live in sand and in reef crevices.

Moray and Conger Eels (Muraenidae and Congridae)

Very little is known about the specific habits of the different moray eel species. They are all predators and feed on a wide variety of fishes, crustaceans and occasionally other invertebrates. During the 1970s, scientists from the University of Hawaii collected moray eels at JA, especially the giant moray, *Gymnothorax javanicus*, in order to extract the marine toxin ciguatera. Ciguatera is a toxin that originates in a microalga (dinoflagellate), which is then eaten by small herbivorous fishes who are then eaten by predators, especially the eels. The toxin accumulates up the food chain. Some of these eels had such high levels of toxin in their tissues that it would be deadly for human consumption. Other big predators such as jacks and barracudas are also frequently ciguateric.

GIANT MORAY, *Gymnothorax javanicus.*

Distribution: Indo-Pacific. Max. size 96 inches (244 cm). *G. Jacanicus* is reportedly the largest of the moray eels and unofficially some ichthyologists have reported individuals exceeding 10 feet (3 m) in length. Other species of moray eels that occur at Johnston include; *Anarchias allardicei, A. cantonensis, A. seychellensis, Echidna leucotaenia, E. polyzona, E. unicolor, Enchelycore pardalis, Gymnomuraena zebra, Gymnothorax berndti, G. buroensis, G. fuscomaculatus, G. gracilicaudus, G. maragitophorous, G. nudivomer, G. nuttingi, G. pindae, G. zonipectis,* an unidentified *Gymnothorax sp., Uropterygius fuscoguttatus, U. inornatus, U. macrocephalus, U. supraforatus,* and *U. tigrinus.*

YELLOWMARGIN MORAY, *Gymnothorax flavimarginatus*. Distribution: Indo-Pacific and tropical eastern Pacific. Max. size 48 inches (122 cm).

UNDULATED MORAY, *Gymnothorax undulatus*. Distribution: Indo-Pacific and tropical eastern Pacific. Max. size 41 inches (104 cm).

TIGER MORAY, *Sculicaria tigrina*. Distribution: Indo-Pacific and tropical eastern Pacific. Max. size 48 inches (122 cm)..

STOUT MORAY, *Gymnothorax eurostus*. Distribution: Indo-Pacific and tropical eastern Pacific. Max. size 23 inches (58 cm).

MUSTACHE CONGER, *Conger cinereus marginatus* (Congridae). Distribution: Indo-Pacific but Hawaiian population is a subspecies. Max. size 54 inches (137 cm).

This eel looks like a moray eel but is actually in a different family. The easiest way to distinguish between the two is that conger eels have pectoral fins and moray eels do not. The conger eel is a nocturnal predator feeding on sleeping fishes and nocturnally active crustaceans. It is rarely seen during the day. The larvae appear like transparent ribbons in the water and attracted to lights at night. The larval form is named leptocephalus and if captured and placed in an aquarium, it will metamorphose into a juvenile within a few days while it stays hidden under gravel or in a small crevice. A second species of conger eel, *Conger oligoporus*, was seen from the sub *Makalii* at 356 meters depth.

WHITEMOUTH MORAY, *Gymnothorax meleagris*. Distribution: Indo-Pacific and Galapagos Islands. Max. size 40 inches (1 m). This eel is very common in the lagoon. It is known to eat small fish and crabs. Its typical behavior is shown in the photo, watching while slightly emerged from a reef crevice. Eels can be both ambush predators and can steathily stalk prey. Sometimes when morays are swimming in the open, they are followed by groupers and jacks.

PEPPERED OR CRAB MORAY, *Siderea picta*. Distribution: Indo-pan-Pacific. Max. size 48 inches (122 cm). This amazing eel hunts along the water edge and in shallow tide pools. It often can capture the sallylightfoot crab (*Grapsus tenucrustatus*) by crawling onto land and grabbing crabs which are overheated and sluggish during the day. This species is widespread in the Pacific but populations are in decline due to human overfishing and habitat destruction. The eel is a favored food by Pacific Islanders and its habits make it easy human prey. Its prime habitat of tidepools is often covered with sand to make more appealing tourist beaches in places like Hawaii.

Series showing eel sneaking up on a crab from behind.

Eel as it opens jaws to bite the crab from behind.

Eel dragging the poor crab into the water where it will be devoured.

Eel hunting crab on a pier piling. The eel often holds its head above water looking for prey. Crabs are generally very wary of these eels.

This eel is often seen along the south shore of Sand Island. It is most active in hunting crabs during low tides on calm days and nights. The eel's typical hunting behavior is to swim slowly along the water's edge and to slither onto wet but emergent rock to get from tidepool to tidepool.

Small Reef Predators

Lizardfish, Trumpet, Cornet and Hawkfish, Goatfish

Large reef predators like jacks and sharks are conspicuous and attract the most attention from divers. But there are also a variety of small and very common predatory fishes that are actually very abundant. These small predatory species feed mainly by ambushing prey and are most successful in capturing naïve juvenile fishes. Most of these small predatory carnivorous fishes are opportunistic feeders and will snap at any tiny animal that ambles blindly nearby..

TWOSPOT LIZARDFISH, *Synodus binotatus* (Synodontidae). Distribution: Indo-Pacific. Max. size 7 inches (18 cm).

SAND LIZARDFISH, *Synodus nebulosa* (Synodontidae). Distribution: Indo-Pacific. Max. size 8 inches (20 cm). There are five species belonging to two genera of lizardfishes on JA. All are cryptic ambush predators and are found usually on sand adjacent to corals. This group comprises 1.7% of the total fish fauna.

DEVIL SCORPIONFISH, *Scorpaenopsis diabolus* (Scorpaenidae). Distribution: Indo-Pacific. Max. size 12 inches (30 cm). Scorpionfish have venomous spines and can inflict painful stings if one accidentally bumps into one. They are camouflaged to match the reef background and ambush small fishes and shrimps that wander too close. There is a total of 11 species (6 genera) of scorpionfish at JA, which comprise about 3.7% of the total fish fauna.

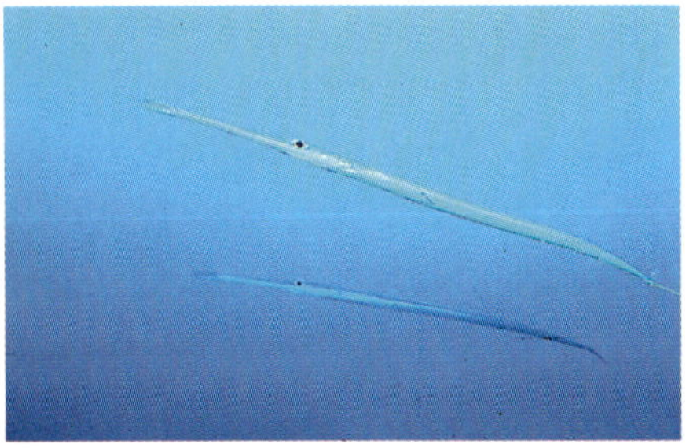

CORNETFISH, *Fistularia commersonii* (Fistularidae). Distribution: Indo-Pacific and tropical eastern Pacific. Max. size 60 inches (152 cm).

This predator is frequently seen in the lagoon often in schools of several to dozens of individuals. Their behavior and coloration makes them very hard to see in the water column especially if the water is murky. They have a similar appearance to the trumpetfish but can be distinguished by the long filament trailing off the tail and by being dorso-ventrally compressed, whereas the trumpetfish is laterally compressed. The long snout of both species functions like a slurp-gun. These fishes slowly approach potential prey and by a rapid opening of their mouth create suction and inhale small prey. There is only this one species of fistularid at JA.

TRUMPETFISH, *Aulostomus chinensis* (Aulostomidae). Dark color phase. Distribution: Indo-Pacific and tropical eastern Pacific. Max. size 30 inches (76 cm). The trumpetfish is a chameleon on the reef and can change its color depending on the situation. There is only this one species of aulostomid at JA.

TRUMPETFISH, *Aulostomus chinensis*. Yellow color phase. This individual is accompanying a school of herbivorous yellow tangs. These tangs are too large for it to swallow. Instead it is using the tangs as camoflage to flush small critters out from cover. As the yellow tangs graze reef algae, small animals hiding in the algal masses are exposed and the trumpetfish attacks.

Flowery flounder, *Bothus mancus* (Bothidae). Distribution: Indo-Pacific and tropical eastern Pacific. Max. size 18 inches (46 cm). Flounders are also ambush predators and can be frequently seen in very shallow water where the smallest fishes are often hovering at the waters edge. Their most common prey in this situation, are very young juvenile mullet. Larger reef fish predators are unable to swim in water only a few inches deep but the flounder is obviously adapted for this situation. In addition to this flatfish, *Samariscus triocellatus* (Pleuronectidae) is also at JA.

Houndfish, *Tylosurus crocodilus* (Belonidae). Distribution: worldwide. Max size 53 inches. (135 cm). The Houndfish and other needlefish hover just beneath the water's surface and feeds on a variety of small swimming fish and crustaceans. This photo is a specimen caught fly-fishing.

Hawkfishes (Cirrhitidae)

These fish are so named for their habit of swooping out from their reef perch to capture small fish and plankton drifting in the water. In addition to the three species illustrated, *Amblycirrhitus bimacula* is also rarely found on JA.

Arc-eye hawkfish, *Paracirrhites arcatus*. Distribution: Indo-Pacific. Max. size 5.5 inches (14 cm). This species occurs in two color patterns and both co-occur in the same reef areas. However, I have never seen the two color morphs intermingle; they are always paired with an individual of the same color pattern. The inset is the dark color phase.

Stocky hawkfish, *Cirrhites pinnulatus*. Distribution: Indo-Pacific. Max. size 12 inches (30 cm).

BLACKSIDE HAWKFISH, *Paracirrhites forsteri*—small adult ~ 4 inches (10 cm)

Paracirrhites forsteri—young juvenile ~2 inches (5 cm).

Paracirrhites forsteri—adult ~8 inches (20 cm).

Paracirrhites forsteri— juvenile ~3 to 4 inches (8 to 10 cm).

Goatfishes (Mullidae)

Goatfish are predators on very small fish, crustaceans and other animals that are usually found buried in sand. They use their specialized chin-barbels, which are covered with taste buds to detect prey hidden in sand. Goatfish eat tiny crabs, worms, shrimps, and ____While digging their prey out of the sand, they often swallow mouthfuls of sand with their food. There is a total of 7 species (2 genera) of goatfish at JA, which comprise about 2.3% of the total fish fauna. These fish are one of the popular fishery species and among those most frequently caught.

Goatfishes are one of our main study species used as a key indicator for assessing possible bioaccumulation of contamination from lagoon sediments, since the goatfishes are predators of small sand-dwelling worms, snails and crustaceans that live in and filter feed sediments directly. Contamination, if present, should be evident through bioaccumulation of contaminants in goatfish tissues.

MANYBAR GOATFISH, *Parupeneus multifasciatus*. Distribution: Central and western Pacific. Max. size 11 inches (28 cm). This species is one of the ten most common species at JA. An estimated population of 61,850 fish live in the lagoon.

Doublebar goatfish, *Parupeneus bifasciatus* —top. Distribution: Indo-Pacific. Max. size 13 inches (33 cm). This species is one of the main fishery species at JA. An estimated population of 48,000 fish live in the lagoon. This species is under taxonomic study by J. Randall and R. Myers, who have determined that this fish is not the same as the *P. bifasciatus* originally described from the Indian Ocean. They will be renaming the Johnston Atoll and Hawaiian one shown here as a new species.

Blue goatfish, *Parupeneus cyclostomus*—bottom. Distribution: Indo-Pacific. Max. size 20 inches (51 cm). This species is one of the main fishery species at JA. An estimated population of 27,600 fish lives in the lagoon. This is the most favored as food fish of the group.

Sidespot goatfish, *Parupeneus pleurostigma*. Distribution: Indo-Pacific. Max. size 13 inches (33 cm). This species is one of the less frequently seen goatfish and is usually found in the central part of the lagoon near the main channel.

YELLOWSTRIPE GOATFISH, *Mulloidichthys flavolineatus*. Distribution: Indo-Pacific. Max. size 16 inches (41 cm). This species usually displays a black spot on its side, below the first dorsal fin. Note the chin barbels extended as the fish searches the sand for invertebrate prey. This species' population has been estimated to be about 188,900 individuals in the lagoon. It is one of the top ten fishery species at JA and the 14th (0.4%) most abundant species overall.

YELLOWFIN GOATFISH, *Mulloidichthys vanicolensis*. Distribution: Indo-Pacific. Max. size 15 inches. Very similar to *M. flavolineatus* but without the black spot on the side. Both species aggregate in resting groups and mostly feed at night.

Nocturnal Fishes (Apogonidae, Holocentridae and Priacanthidae)

Iridescent cardinalfish, *Apogon kallopterus* (Apogonidae). Distribution: Indo-Pacific. Max. size 6 inches (15 cm).

This group of fishes is seen during the daytime hovering under coral ledges and in the caves on the reef. At night, they migrate out into open water and hunt for zooplankton. There are 6 species in 4 genera of cardinalfish (Apogonidae), 13 species in 5 genera of squirrelfish (Holocentridae) and 3 species in 2 genera of glasseyes (Priacanthidae). The species illustrated are the most common ones seen by divers on the reef. Other species of cardinalfish not shown include; *Apogon erythrinus, A. taeniopterus, Apogonichthys perdix, Epigonus sp.*, and *Pseudamiops gracilicauda*. Other species of squirrelfish not shown include; *Myripristis chryseres, M. kuntee, Neoniphon aurolineatus, N. sammara, Plectrypops lima, Pristilepis oligolepis, Sargocentron microstoma, and S. punctatissimum*. Other species of glasseyes not shown include; *Cookeolus boops* and *Pricanthus meeki.*

Saber squirrelfish, *Sargocentron spiniferum* (Holocentridae). Distribution: Indo-Pacific. Max. size 18 inches (46 cm).

TAHITIAN SQUIRRELFISH, *Sargocentron tiere* (Holocentridae). Distribution: Indo-Pacific. Max. size 13 inches.

HAWAIIAN SQUIRRELFISH, *Sargocentron xantherythrum* (Holocentridae). Distribution: Hawaii and Johnston Atoll endemic. Max. size 6.5 inches (16 cm).

Brick soldierfish, *Myripristis amaena* (Holocentridae). Distribution: Oceania. Max. size 10 inches (25 cm). This species is one of the most favored food fishes of Hawaiians and others and it is the number one fish to be caught in the JA recreational fisheries. In a resource inventory conducted by Irons et al 1990, it was estimated that the population of this species in the lagoon was about 385,400 individuals. The total number caught in the local fisheries was about 3362, which is less than one percent, and probably has insignificant population impact (see also Dee and Parrish 1994).

Bigscale soldierfish, *Myripristis berndti* (Holocentridae). Distribution: East Africa to the tropical eastern Pacific. Max. size 11 inches (28 cm).

GLASSEYE, *Heteropriacanthus cruentatus* (Priacanthidae). Distribution: Circumtropical. Max. size 13 inches (33 cm). This is another nocturnal fish that is found hiding under coral ledges during daytime. It emerges when dark to feed on large zooplankton.

Butterflyfishes (Chaetodontidae)

Butterflyfishes are very similar to one another in morphology but differ radically in color patterns. Fish see color is differently than humans. Dr. George Losey, University of Hawaii, has determined that many reef fishes can see ultraviolet wavelengths and that their color patterns appear quite different as a result. Also, at the University of Hawaii, Dr. Ernie Reese has developed methods of monitoring reef community health and status by conducting underwater surveys of butterflyfish species distribution and abundances. There are 18 species in 4 genera of butterflyfish at Johnston Atoll and comprise about 6% of the total fish fauna. Butterflyfish are carnivores and various species feed on live coral, small benthic invertebrates and plankton. The species not shown are *Hemitaurichthys thompsoni* and a deep water species, *Chaetodon modestus.*

Butterflyfishes are interesting to watch because of their social and feeding behaviors. Fish are usually with a long term mate, together they defined a home feeding range. Spawning occurs at dusk. They feed throughout the day by browsing various corals, algae, sponges or zooplankton.

THREADFIN BUTTERFLYFISH, *Chaetodon auriga*. Distribution: Indo-Pacific. Max. size 8 inches (20 cm). This fish feeds on worms and coelenterates.

RACCOON BUTTERFLYFISH, *Chaetodon lunula*. Distribution: Indo-Pacific. Max. size 8 inches (20 cm). Feeds on a variety of tiny invertebrates.

LINED BUTTERFLYFISH, *Chaetodon lineolatus*. Distribution: Indo-Pacific. Max. size 12 inches (30 cm). This is the largest butterflyfish and feeds on coral polyps and anemones.

KEYHOLE BUTTERFLYFISH, *Chaetodon unimaculatus*. Distribution: Indo-Pacific. Max. size 8 inches (20 cm). Feeds on a variety of tiny invertebrates and corals.

SADDLEBACK BUTTERFLYFISH, *Chaetodon ephippium*. Distribution: Central and western Pacific. Max. size 9 inches (23 cm). Feeds on a variety of tiny invertebrates.

Milletseed butterflyfish, *Chaetodon miliaris*. Distribution: Hawaii and Johnston Atoll endemic. Max. size 6 inches (15 cm). It swims high above the reef substrate and feeds on zooplankton. Rare at Johnston.

Speckled butterflyfish, *Chaetodon citrinellus*. Distribution: Indo-Pacific. Max. size 5 inches (13 cm). Feeds on a variety of tiny invertebrates and zooplankton.

Multiband butterflyfish, *Chaetodon multicinctus*. Distribution: Hawaii and Johnston Atoll endemic. Max. size 5 inches (13 cm). Feeds almost exclusively on live coral polyps.

RETICULATED BUTTERFLYFISH, *Chaetodon reticulatus*. Distribution: Central and western Pacific. Max. size 7 inches (18 cm). Feeds almost exclusively on live coral polyps. This species is usually found on the far eastern reefs of the atoll.

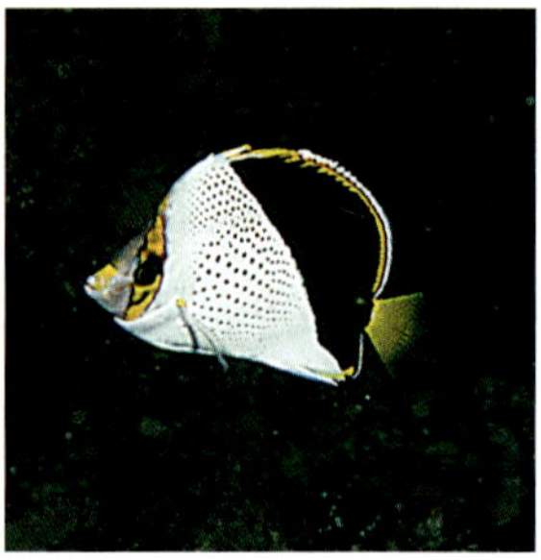

TINKER'S BUTTERFLYFISH, *Chaetodon tinkeri*. Distribution: Hawaii, Johnston Atoll and Marshall Islands. Max. size 7 inches (18 cm). This is a deepwater species and rarely is seen shallower than 75 feet. Feeds on a variety of tiny invertebrates.

FOURSPOT BUTTERFLYFISH, *Chaetodon quadrimaculatus*. Distribution: Oceania. Max. size 6 inches (15 cm). Feeds on a variety of tiny invertebrates and corals.

ORNATE BUTTERFLYFISH, *Chaetodon ornatissimus*. Distribution: Indo-Pacific. Max. size 8 inches (20 cm). This species is an obligate coral feeder and does not survive in aquaria. This fish is usually found in mated pairs. Their long term pair bond can last for years during which time they remain in the same reef area.

CHEVRON BUTTERFLYFISH, *Chaetodon trifascialis*. Distribution: Indo-Pacific. Max. size 7 inches (18 cm). Common at Johnston but rare in Hawaii. This fish is the 4th (9%) most abundant species overall in the lagoon. It aggressively maintains its territory which usually is a tabletop *Acropora cytherea* from neighboring conspecifics. It feeds almost exclusively on the coral it defends. The juvenile is shown in the inset.

OVAL BUTTERFLYFISH, *Chaetodon lunulatus*. Distribution: Central and western Pacific. Max. size 5.5 inches (14 cm). This species is also an obligate coral feeder.

LONGNOSE BUTTERFLYFISH, *Forcipiger flavissimus*. Distribution: Indo-Pacific and tropical eastern Pacific. Max. size 7 inches (18 cm). This fish uses its long snout to get tiny crustaceans hidden in reef nooks and crannies. It also pecks between the spines of sea urchins to eat the tube feet.

PENNANTFISH, *Heniochus diphreutes*. Distribution: Indo-Pacific. Max. size 8 inches (20 cm). This fish swims in schools high above the reef substrate and feeds on zooplankton.

Angelfishes (Pomacanthidae)

Angelfishes have a similar appearance to butterflyfishes and are most easily distinguished by the spine on the gill cover (opercle) of the anglefishes. There is a total of 8 anglefish species (3 genera) at JA which comprise about 2.7% of the total fish fauna. One species is endemic to Johnston Atoll. Another species, in the genus Genicanthus, has been seen from the UH submersible but never collected. *Centropyge nigriocellus* is also listed as occurring at Johnston but I have never seen it.

RAINBOW PYGMY ANGELFISH, *Centropyge nahackyi*. Distribution: Johnston Atoll endemic. Max. size 5 inches (13 cm). One specimen has been found in Hawaii but the only breeding size population is known at JA. This fish is the one species that is clearly recognized as unique to JA. It has been seen picking at sponges and small reef invertebrates. It is a deepwater species and is associated with caves and ledges below 60 feet (18 m).

MULTICOLORED PYGMY ANGELFISH, *Centropyge multicolor*. Distribution: Western Pacific. Max. size 4 inches (10 cm). Only one specimen has been found at Johnston Atoll.

BROWN PYGMY ANGELFISH, *Centropyge fisheri*. Distribution: Hawaii and Johnston Atoll endemic. Max. size 3 inches (8 cm). Sightings at Johnston are uncertain and a specimen needs to be obtained. It is herbivorous.

Potter's pygmy angelfish, *Centropyge potteri*—female. Distribution: Hawaii and Johnston Atoll endemic. Max. size 5 inches (13 cm). This fish is herbivorous. In habitats where many males can defend territories, the fish is usually in pairs but on patch reefs where one male can become dominant, that male gets a harem.

Potter's pygmy angelfish, *Centropyge potteri*—male.

FLAME PYGMY ANGELFISH, *Centropyge loricula*. Distribution: Oceania. Max. size 4 inches (10 cm). This species is the most common angelfish at JA. It is herbivorous.

BANDIT ANGELFISH, *Desmoholocanthus arcuatus*. Distribution: Hawaii and Johnston Atoll endemic. Max. size 7 inches (18 cm). This deepwater species feeds mainly on sponges.

Damselfishes (Pomacentridae)

Pomacentrids are the 11th most abundant fish family overall in the lagoon. There is a total of 9 damselfish species (4 genera) at JA, which comprise about 3% of the total fish fauna. The most curious phenomenon at JA is the total absence of one type of damselfish. Completely missing are species that farm algae and create lush algal lawns. The Pacific gregory, *Stegastes fasciolatus* occurs in Hawaii and other, closely related species are found in the Line Islands but none occur at JA. This has possibly had a profound impact on the reef ecosystem due to the fact that damselfish territories with lush algal farms harbor a large mass of nitrogen fixing algae and also create a refuge for many tiny invertebrates. The lack of the guild of territorial algal farming damselfishes at JA may have consequences for reef diversity. This has been a longtime topic in our atoll research. The one species not picturedis *Plectroglyphidodon phoenixensis.*

HAWAIIAN SERGEANT MAJOR, *Abudefduf abdominalis*. Distribution: Hawaii and Johnston Atoll endemic. Max. size 10 inches (25 cm). This species was not found at JA until recently. Several individuals were seen for a few years off North Island but this population disappeared after four years. In 1998, hundreds of this species and its close relative, *A. vaigiensis* appeared on JA reefs and established breeding populations. Current research is focused on how the atoll ecology may be affected by the recent colonization of these two species. Feeds mainly on zooplankton.

Indo-Pacific sergeant major, *Abudefduf vaigiensis*. Distribution: Indo-Pacific. Max. size 8 inches (20 cm). This species recently appeared on JA for the first time. Individuals are now abundant and breeding. Feeds mainly on zooplankton. This species co-occurs and potentially hybridizes with *A. abdominalis* on JA.

Blackspot sergeant major, *Abudefduf sordidus*—female. Distribution: Indo-Pacific. Max. size 9.3 inches (24 cm). A benthic omnivore eating a variety of plants and small invertebrates.

Blackspot sergeant major, *Abudefduf sordidus*—male breeding coloration. This species produces a huge mass of eggs every several days, which are guarded by the male until they hatch after 5 or 6 days. As part of her Ph.D. research, Lisa Kerr Lobel has shown how this species can be used as an indicator of contaminants in the marine environment. She has developed a method using the incidence of embryo abnormalities as a measure of contaminant effects.

Agile chromis, *Chromis agilis*. Distribution: Indo-Pacific. Max. size 4 inches (10 cm). Feeds mainly on zooplankton. This species is found on shallow reefs.

Threespot chromis, *Chromis verater*. Distribution: Hawaii and Johnston Atoll endemic. Max. size 8.5 inches (22 cm). Feeds mainly on zooplankton. This species inhabits mainly deepwater reefs.

Bluefin chromis, *Chromis acares*. Distribution: Central and western Pacific. Max. size 2 inches (5 cm). Common at Johnston but rare in Hawaii. Feeds mainly on zooplankton.

Brighteye damselfish, *Plectroglyphidodon imparipennis*. Distribution: Indo-Pacific. Max. size 2.5 inches (6 cm). Feeds on a variety of benthic invertebrates and zooplankton.

Blueeye damselfish, *Plectroglyphidodon johnstonianus* . Distribution: Indo-Pacific. Max. size 5 inches (13 cm). This fish is the 15th (0.4%) most abundant species overall in the lagoon. Feeds on coral polyps.

Domino damselfish, *Dascyllus albisella*. Distribution: Hawaii and Johnston Atoll endemic. Max. size 5 inches (13 cm). Feeds mainly on zooplankton. (Juvenile in inset.)

Domino damselfish, *Dascyllus albisella* aggregation in antler coral *Pocilopora eydouxi*. This species is conspicuous because it produces loud courtship sounds that divers can easily hear. This species has been the research subject of two Ph.D. dissertations (by David Mann and Steven Oliver) conducted at JA. These studies investigated the role of acoustic communication in the mating habits of this fish. It was found that the male who constantly made vigorous sounds more frequently also had the highest mating success.

Labroid Fishes

Wrasses (Labridae) and Parrotfishes (Scaridae)

This group of fishes is classified as two families, which are merely separated by their type of food habit. Combined together, the labroid fishes are the dominant group on the reef in terms of biomass and number of species (total 35 species, 21 genera). Wrasses are carnivorous, while parrotfishes are mostly herbivorous (although some species feed on corals also). This group has a unique morphological architecture of bones and muscles in their throat that enables these fish to thoroughly grind all ingested foods. Most other fishes can only swallow prey whole and then have to rely on digestion to break down food material.

Wrasses (Labridae)

There is a total of 30 wrasse species (belonging to 19 genera) at JA, which comprise about 9.6% of the total fish fauna. It is the second most speciose family behind the moray eels. Wrasses are carnivores and will take most animal food that they can capture. Those species not pictured include; *Anampses cuvier, Cirrhilabrus jordani, C. luteovittatus, Coris ballieui, Pseudocheilinus evanidus, P. tetrataenia, P. ocellatus, Pseudojuloides cerasinus, Wetmorella albofasciatia, Xyrichtys pavo* and two deep water species *Polylepion russelli and Xyrichtys aneitensis.*

CLEANER WRASSE, *Labroides phthirophagus* (fish with yellow stripe on the right in photo). Distribution: Hawaii and Johnston Atoll endemic. Max. size 4 inches (10 cm). There appear to be slight color differences between the JA and Hawaiian populations. This is the most specialized wrasse in terms of feeding habits. It picks the paprasites off the sides of other fishes. These other fishes learned to be cleaned and swim and pose in front of cleaner wrasses to get their attention.

CLEANER WRASSE, *Labroides phthirophagus* cleaning the parrot fish *Calatomus carolinus* (Scaridae). The cleaner tickles its host fish while inspecting for parasites to eat.

HAWAIIAN HOGFISH, *Bodianus bilunulatus albotaeniatus*. Distribution: Indo-Pacific with a Hawaiian subpopulation. Max. size 20 inches (51 cm). (Juvenile in inset.). This wrasse eats any small fish it can get, but mainly preys upon benthic crustaceans and small urchins.

RINGTAIL WRASSE, *Oxycheilinus unifasciatus*. Distribution: Central and western Pacific. Max. size 18 inches (46 cm). (Juvenile in inset.)

CLOWN OR YELLOWTAIL WRASSE, *Coris gaimard*—supermale. Distribution: Central and western Pacific, eastern Indian Ocean. Max. size 15 inches (38 cm).

Clown or yellowtail wrasse, *Coris gaimard*—initial phase male or female. (Small juvenile in inset,)

Cigar wrasse, *Chelio inermis*—female. Distribution: Indo-Pacific. Max. size 20 inches (51 cm).

Cigar wrasse, *Chelio inermis*—male

Slingjaw wrasse, *Epibulis insidiator*—initial phase on left, yellow phase (often female) on right, supermale below. Distribution: Indo-Pacific. Max. size 12 inches (30 cm). Common at Johnston but rare in Hawaii.

Bird wrasse, *Gomphosus varius*—immature on left, very small juvenile on right, and female in inset. Distribution: Central and western Pacific. Max. size 12 inches (30 cm). This fish is the 11th most abundant species overall in the lagoon and comprises about 0.8% of all species.

Bird wrasse, *Gomphosus varius*—male. The long snout enables plucking small crustaceans from tiny crevices.

Ornate wrasse, *Halichoeres ornatissimus* —planktonic larva settlement stage on left, juvenile on right, initial phase below. Distribution: Central and western Pacific. Max. size 6 inches (15 cm).

ORNATE WRASSE, *Halichoeres ornatissimus*—supermale (terminal phase male).

EIGHT STRIPE WRASSE, *Pseudocheilinus octotaenia*. Distribution: Indo-Pacific. Max. size 5 inches (13 cm).

SHORT-NOSE WRASSE, *Macropharyngodon geoffroy*. Distribution: Hawaii and Johnston Atoll endemic. Max. size 6 inches (15 cm).

ROCK MOVER, *Novaculichthys taeniourus*. Distribution: Indo-Pacific and tropical eastern Pacific. Max. size 12 inches (30 cm). (Juvenile in inset.)

Hybrid Saddle wrasse, *Thalassoma duperrey X lutescens*—hybrid. Distribution: Johnston Atoll endemic. Max. size 10 inches (25 cm). The hybrid ecology and behavior is under study.

Saddle wrasse, *Thalassoma duperrey* . Distribution: Hawaii and Johnston Atoll endemic. Max. size 10 inches (25 cm). this species (including the hybrid) is the sixth most common lagoon fish with and comprises an estimated 3,6% of the reef fish population. (Juvenile in inset.)

Sunset wrasse, *Thalassoma lutescens*—adult female. Distribution: Indo-Pacific. Max. size 10 inches (25 cm). This species is very common on Pacific islands and is seen not uncommonly on Johnston Atoll, but is very rare in Hawaii.

Sunset wrasse, *Thalassoma lutescens*—adult supermale.

BLACKTAIL WRASSE, *Thalassoma ballieui*. Distribution: Hawaii and Johnston Atoll endemic. Max. size 16 inches (41 cm).

SURGE WRASSE, *Thalassoma purpureum*—terminal phase male. Distribution: Indo-Pacific. Max. size 17 inches (43 cm). This specimen appears to have an injury above the eye.

CHRISTMAS WRASSE, *Thalassoma trilobatum*—male. Distribution: Indo-Pacific. Max. size 12 inches (30 cm).

FIVE-STRIPED WRASSE, *Thalassoma quinquevittatum*—initial phase male. Distribution: Indo-Pacific. Max. size 7 inches (18 cm). Females are similar in color to these initial phase males.

FIVE-STRIPED WRASSE, *Thalassoma quinquevittatum*—terminal phase male.

Belted wrasse, *Stethojulis balteata*—male. Distribution: Hawaii and Johnston Atoll endemic. Max. size 6 inches (15 cm).

Belted wrasse, *Stethojulis balteata*—female. Small wrasses eat a variety of tiny animals that hide among coral, algae, and sand. Prey items include mollusks, amphipods, copepods, worms, crabs, brittlestars, fish eggs, and tiny fish.

Parrotfishes (Scaridae)

There is a total of 6 parrotfish species (2 genera) at JA which comprise about 2% of the total fish fauna. One species, *Scarus sordidus*, is the second most abundant species in the atoll and comprises about 16.2% of the overall reef fish population. The parrotfish are all herbivores and use their strong beak-like jaws to rasp dead coral for ingrown algae. Species not pictured include *Calatomus carolinus, Scarus dubius*, and *Scarus rubroviolaceus*.

BULLETHEAD PARROTFISH, *Chlorurus sordidus* —juveniles. Distribution: Indo-Pacific. Max. size 16 inches (41 cm).

BULLETHEAD PARROTFISH, *Chlorurus sordidus*—female.

BULLETHEAD PARROTFISH, *Chlorurus sordidus*—initial phase male.

Male *Chlorurus sordidus* biting into dead coral to get at the overgrown and ingrown algae. In the inset are parrotfish bite marks scraped into a lush layer of microalgae (red alga, *Polysiphonia*) overgrowing dead coral. The bite marks leave clear space onto which other algae, corals, sponges or other benthic encrusting organisms can attach and grow.

Adult *Chlorurus sordidus* female. Note the filamentous algae (possibly a blue-green) growing on teeth on side of mouth. The algae is actually attached and growing into the fish's bone. How this happens or whether it is detrimental to the fish is unknown.

Speckled parrotfish, *Chlorurus perspicillatus*—female. Distribution: Hawaii and Johnston Atoll endemic. Max. size 24 inches (61 cm).

Speckled parrotfish, *Chlorurus perspicillatus*—male. This species is one of the top ten fishery species and has an estimated population size of 29,450 individuals in the lagoon.

Palenose parrotfish, *Scarus psittacus* —female. Distribution: Indo-Pacific. Max. size 12 inches (30 cm). This fish is the 8th most abundant species overall in the lagoon and comprises about 2.8% of all the fishes.

Palenose parrotfish, *Scarus psittacus* —male.

Surgeonfishes (Acanthuridae)

There is a total of 21 surgeonfishes species (4 genera) at JA which comprise about 7% of the total fish fauna. It is the third dominant fish family on the reefs and two of the species are important in the local fisheries. All of the surgeonfishes are herbivorous but vary in the kinds of algae they consume and in whether they ingest sand too. Species with thick-wall stomachs and ingest fine grain sand with algae are called grazers. Species that avoid sand ingestion and have thin-wall acidic stomachs are browsers. Species not pictured include *Acanthurus guttatus, Acanthurus nigricauda, Ctenochaetus marginatus, Naso hexacanthus, Naso unicornis, Naso brevirostris*, and an unidentified *Naso* species. All surgeonfishes have distinct spines near the tail.

GOLDRING SURGEONFISH, *Ctenochaetus strigosus*. Distribution: Indo-Pacific. Max. size 7 inches (18 cm). This species exhibits population differences between different island groups, suggesting that there is a high degree of local isolation, even between Hawaii and JA. This fish is the number one most abundant species overall in the lagoon and comprises an estimated 24.6% of the total fish population. The estimated population size is 1,650,300 individuals. It is also one of the top two fishes taken by fisherman on the atoll with a typical annual harvest of about 1,200 fish.

This species is an herbivore grazer feeding mainly on microalgae mixed thickly with fine grain sand particles. This species is also often associated with Ciguatera. It digests algal food mainly by mechanical trituration in a thick-walled stomach. Sand is processed through the gut along with food. Because this fish directly consumes and digestively processes sand it is a key species for monitoring uptake of contaminants found in the sediments.

This photo shows a mass of young surgeonfishes that have recently migrated from the open ocean where they developed as larvae. Once on the reef they metamorphose into juvenile fish. During this metamorphosis,they change from being transparent and silvery to dark colored and will soon develop the coloration typical for their species. It is during this transitional phase, that these little fishes are particularly vulnerable to reef predators. The numbers will soon be decimated by predation and only a fraction of these fish will survive to adult. The majority of fish shown in this photo are juvenile *Ctenochaetus strigosus* and *Acanthurus triostegus*. The fish are approximately 1inch length (2.5cm) and they have spent about 2 months in the open ocean as larvae and have been on the reef for between two to five days.

ACHILLES TANG, *Acanthurus achilles*. Distribution: Oceania. Max. size 10 inches (25 cm). This fish is the 5th most abundant species overall in the lagoon and comprises about 5.8% of all the fishes.

GOLDRIM SURGEONFISH, *Acanthurus nigricans*. Distribution: Indo-Pacific to eastern Pacific. Max. size 8 inches (20 cm). This fish is the 13th most abundant species overall in the lagoon and comprises about 0.6% of all the fishes.

RINGTAIL SURGEONFISH, *Acanthurus blochii*. Distribution: Indo-Pacific. Max. size 17 inches (43 cm).

WHITEBAR SURGEONFISH, *Acanthurus leucopareius*. Distribution: Anti-equatorial, Hawaii to southern Japan in the north and Easter Island to New Caledonia in the south. Max. size 10 inches (25 cm).

BLUE-LINED SURGEONFISH, *Acanthurus nigroris*. Distribution: Oceania. Max. size 10 inches (25 cm). This fish is the 3rd most abundant species overall in the lagoon and comprises about 10.4% of all the fishes.

EYE-STRIP SURGEONFISH, *Acanthurus dussumieri*. Distribution: Indo-Pacific. Max. size 18 inches (46 cm). Surgeonfishes often forage in groups consisting of several species. These mixed species feeding aggregations move as a herd across the reef. A main target is the lush algal garden in damselfish territories. The herd is able to overwhelm the territorial damselfishes. However, at Johnston Atoll, alga farming damselfish (*Stegastes* spp) do not occur, thus surgeonfish feeding aggregations rarely occur.

ORANGEBAND SURGEONFISH, *Acanthurus olivaceus*. Distribution: Oceania and western Pacific. Max. size 12 inches (30 cm). (Juvenile in inset.)

THOMPSON'S SURGEONFISH, *Acanthurus thompsoni*. Distribution: Indo-Pacific. Max. size 11 inches (28 cm).

ORANGESPINE UNICORNFISH, *Naso lituratus*. Distribution: Indo-Pacific. Max. size 18 inches (46 cm). This fish is the 12th most abundant species overall in the lagoon and comprises about 0.8% of all the fishes.

CONVICT SURGEONFISH (MANINI), *Acanthurus triostegus sandvicensis.* Distribution: Hawaii and Johnston Atoll endemic subspecies. Acanthurus triostegus is found throughout the Indo-Pacific and tropical eastern Pacific. Max. size 11 inches (28 cm). An herbivore browser feeding mainly on fine filamentous algae while avoiding ingestion of carbonate sand particles. It digests alga food mainly by acid-lysis in a thin-walled and distensible stomach. Hawaiian and JA specimens have a long sickle shape black color streak at base of pectoral fin. This species is one of the top ten fishery species, has an estimated population size of 599,600 individuals in the lagoon and comprises about 1.6% of the total fish population making it the tenth most abundant fish.

This photo shows *Acanthurus triostegus* from relatively nearby Fanning Atoll, Line Islands. It has a small spot at the base of pectoral fin. Convict surgeonfish at other pacific atolls have different pectoral fin spot patterns and populations are readily distinguishable.

BLACK SURGEONFISH, *Ctenochaetus hawaiiensis*. Distribution: Oceania. Max. size 11 inches (28 cm). Upper left–small juvenile ~1 inch (2.5 cm). Upper right —medium juvenile ~2 inches (5 cm). Middle—medium juvenile ~2 inches (5 cm) Bottom— adult ~6 to 10 inches (15 to 25 cm).

SAILFIN TANG, *Zebrasoma veliferum*. Distribution: Central and western Pacific. Max. size 15 inches (38 cm). (Juvenile in inset.)

YELLOW TANG, *Zebrasoma flavescens*. Distribution: North Pacific from Hawaii to southern Japan but abundant only in Hawaii and Johnston Atoll. Max. size 8 inches (20 cm). This fish is the 7th most abundant species overall in the lagoon and comprises about 3.4% of all the fishes.

Moorish Idol (Zanclidae)

Moorish Idols are closely related to surgeonfishes (previous section) and sea chubs (following section) and are found in the suborder Acanthoidei. These fishes are mostly herbivorous, except the Moorish Idol, which eats sponges. (Surgeonfishes can be identified by having a distinct spine (up to 3) on the base of the tailfin. This easily distinguishes them from other deep bodied colorful reef fish, such as butterflyfishes.)

Moorish Idol, *Zanclus cornutus*. Distribution: Indo-Pacific and tropical eastern Pacific. Max. size 8 inches (20 cm). This genus contains only the one species. Feeds on sponges. Even though this is a strikingly beautiful and conspicuous fish, very little is actually known about its behavior and ecology.

Sea Chubs (Kyphosidae)

Sea Chubs are herbivorous. There are three species on JA and it is important in the local fishery.

Gray chub, *Kyphosus bigibbus*. Distribution: Indo-Pacific. Max. size 24 inches (61 cm).

Highfin chub, *Kyphosus cinerascens*. Distribution: Indo-Pacific. Max. size 20 inches (51 cm).

LOWFIN CHUB, *Kyphosus vaigiensis.* Distribution: Indo-Pacific. Max. size 24 inches (61 cm). This species is one of the top ten fishery species and has an estimated population size of 22,350 individuals in the lagoon.

Bottom Dwellers

Blennies and Gobies

Blennies (Blenniidae)

There is a total of 4 blenny species (3 genera) at JA which comprise about 1.3% of the total fish fauna. The species not shown are *Cirripectes variolosus* and *Istiblennius gibbifrons*. These fishes feed on algae, coral and detritus.

SCARFACE BLENNY, *Cirripectes vanderbilti* (Blenniidae). Distribution: Hawaii and Johnston Atoll endemic. Max. size 4.5 inches (11 cm). Closely related to the Indo-Pacific *C. variolosus* and both occur at Johnston Atoll. Feeds on algae.

SHORT-BODIED BLENNY, *Exallias brevis*—female. Distribution: Indo-Pacific. Max. size 6 inches (15 cm). Feeds on coral polyps

SHORT-BODIED BLENNY, *Exallias brevis*—male.

Gobies (Gobiidae)

There is a total of 8 goby species (5 genera) at JA, which comprise about 2% of the total fish fauna. Those species not shown include *Bathygobius cocosensis, Eviota epiphanes, Gnatholepis anjerensis, Priolepis aureoviridis, Priolepis farcimen*, and *Trimma unisquamis.*

YELLOW CAVE GOBY, *Trimma taylori.* Distribution: Indo-Pacific. Max. size 1 inch (2.5 cm). Feeds on microzooplankton. This species is only found in caves at depths below 75 feet. It hovers upside down along the cave roof and feeds on tiny zooplankton. This species was discovered in Hawaii but has since been found throughout the tropical Pacific.

Coryphopterus sp. . Distribution: Johnston Atoll (possibly Hawaii, and the Ryukyu Islands if it is the same as the unidentified Hawaiian *Coryphopterus*). Max. size 2.5 inches (6 cm). This species was formerly misidentified as *Fusigobius neophytus*, it has yet to be properly identified. Feeds on tiny invertebrates.

Long Body Benthic Fishes

FLAG-TAIL TILEFISH, *Malacanthus brevirostris* (Malacanthidae). Distribution: Indo-Pacific and tropical eastern Pacific. Max. size 12 inches (30 cm). Feeds on tiny benthic invertebrates.

FIRE DARTFISH, *Nemateleotris magnifica* (Microdesmidae). Distribution: Indo-Pacific. Max. size 3 inches (8 cm). Feeds on zooplankton and tiny benthic invertebrates. At one time this fish was included in the goby family but was later found to be distinctly different.

SAND DIVER, *Parapercis schauinslandii* (Pinguipedidae). Distribution: Indo-Pacific. Max. size 5 inches (13 cm). Feeds on zooplankton and tiny benthic invertebrates.

Triggerfishes, Filefish, Boxfishes and Puffers (Balistidae)

There is a total of 6 triggerfish species (4 genera) at JA, which comprise about 2% of the total fish fauna. They mainly eat benthic invertebrates, but a few species feed in the water column on zooplankton.

LAGOON TRIGGERFISH, *Rhinecanthus aculeatus*. Distribution: Indo-Pacific and tropical eastern south Atlantic. Max. size 12 inches (30 cm). Feeds on benthic invertebrates.

LEI TRIGGERFISH, *Sufflamen bursa*. Distribution: Indo-Pacific. Max. size 8.5 inches (22 cm). Feeds on benthic invertebrates.

Bridled triggerfish, *Sufflamen fraenatus*. Distribution: Indo-Pacific. Max. size 15 inches (38 cm). Feeds on benthic invertebrates.

Gilded triggerfish, *Xanthichthys auromarginatus*. Distribution: Indo-Pacific. Max. size 8 inches (20 cm). Feeds on zooplankton.

Black durgon, *Melichthys niger*. Distribution: Circumtropical. Max. size 12 inches (31 cm). This fish is the 9th most abundant species overall in the lagoon and comprises about 2.4% of all the fishes. Feeds on algae and zooplankton.

PINK-TAIL DURGON, *Melichthys vidua*. Distribution: Indo-Pacific. Max. size 13 inches (33 cm). Feeds on benthic invertebrates.

Filefishes (Monocanthidae)

There is a total of 5 filefish species (3 genera) at JA which comprise about 1.7% of the total fish fauna. The one species not shown is *Aluterus scriptus*. The filefishes feed on a variety of benthic invertebrates including tunicates, sponges, crustaceans and worms.

BARRED FILEFISH, *Cantherhines dumerilii*. Distribution: Indo-Pacific and tropical eastern Pacific. Max. size 15 inches (38 cm).

SQUARETAIL FILEFISH, *Cantherhines sandwichiensis*. Distribution: Hawaii and Johnston Atoll endemic. Max. size 7.5 inches (19 cm).

YELLOWTAIL FILEFISH, *Pervagor aspricaudus*. Distribution: Anti-equatorial in the Indo-Pacific. Max. size 4.5 inches (12 cm).

FANTAIL FILEFISH, *Pervagor spilosoma*. Distribution: Hawaii and Johnston Atoll endemic. Max. size 7 inches (18 cm).

Boxfishes (Ostraciidae)

Two species commonly are found on JA reefs. A third species (*Ostracion cubicus*) was reported once in 1954 but never seen since. Boxfishes feed on benthic invertebrates.

SPOTTED BOXFISH, *Ostracion meleagris camurum* —female. Distribution: Hawaii and Johnston Atoll endemic. Max. size 6 inches (15 cm)

SPOTTED BOXFISH, *Ostracion meleagris*—male (phase with gold spots on side). This color phase is typical of the ones in the Indo-Pacific outside of Hawaii. The inset shows a different color phase (without gold spots on the side), which occurs in Hawaii and is considered a separate subspecies (*Ostracion meleagris camurum*). Both color forms co-occur at JA, documentation by Lobel & Randall, in preparation.

WHITLEY'S TRUNKFISH, *Ostracion whitleyi*—female. Distribution: Hawaiian Islands to French Polynesia. Max. size 6 inches (15 cm).

WHITLEY'S TRUNKFISH, *Ostracion whitleyi*—male in breeding coloration..

Puffers (Tetraodontidae)

Puffers range in size from a few inches to about 15 inches in length. There is a total of 5 puffer species (2 genera) at JA. The species not shown include *Arothron hispidus* and *Canthigaster inframacula* which is a deep water species. Puffers feed on a wide variety of benthic invertebrates.

CROWN TOBY, *Canthigaster coronata*. Distribution: Indo-Pacific. Max. size 5 inches (13 cm).

HAWAIIAN WHITESPOTTED TOBY, *Canthigaster jactator*. Distribution: Hawaii and Johnston Atoll endemic. Max. size 3.5 inches (9 cm).

Spotted puffer, *Arothron meleagris*. Distribution: Indo-Pacific and tropical eastern Pacific. Max. size 13 inches (33 cm).

Porcupinefish (Diodontidae)

Only one species in this family has been found at JA.

Porcupinefish, *Diodon hystrix*. Distribution: Circumtropical. Max. size 28 inches (71 cm). This large fish feeds on the bigger benthic invertebrates including gastropods, crabs, and urchins.

Species List

References

Amerson, A. B. and P. C. Shelton 1976 The natural history of Johnston Atoll, Central Pacific Ocean. *Atoll Research Bulletin* 192:1-479.

Anderson, D. M. and P. S. Lobel. 1987 The continuing enigma of ciguatera. *Biological Bulletin* 172: 89-107.

Bentis, C. J., L. Kaufman, S. Golubic. 2000. Endolithic fungi in reef-building corals (Order: Scleractina) are common, cosmopolitan and potentially pathogenic. *Biological Bulletin* 198(2):254-260.

Brock V., R. S. Jones and P. Helfrich 1965 An ecological reconnaissance on Johnston Island and the effects of dredging. *Technical. Report. 5*, 90p, University of Hawaii, Honolulu, HI.

Brock V., W. Van Heukelem and P. Helfrich 1966 An ecological reconnaissance on Johnston Island and the effects of dredging. *Technical. Report. 11*, 56p, University of Hawaii, Honolulu, HI.

Cohen, A., P. S. Lobel, and G. L. Tomasky. 1997 An unusual event of Coral bleaching on Johnston Atoll, Central Pacific Ocean. *Biological Bulletin* 193:276-279.

Cohen. A. L., D. I. Grahm, S. R. Hart & P. S. Lobel. 2001. Kinematic contraol of skeletal Sr/Ca in a symbiotic coral: Implications for the proxy temperature record. *Paleoceanography* 16: 20-26.

Dee A.J., Parrish J.D. 1994. Reproductive and trophic ecology of the soldierfish, *Myripristis amaena*, in tropical fisheries. *Fishery Bulletin* 92:516-530.

Economakis, A. E. and P. S. Lobel 1998 Aggregations of grey reef sharks, *Carcharhinus amblyrhynchos*, at Johnston Atoll, Central Pacific Ocean. *Environmental Biology of Fishes* 51(2):129-139.

Grigg, R. W., 1981. *Acropora* in Hawaii. Part 2. Zoogeography. *Pac. Sci.* 35:(1):15-24.

Grigg, R. W., J. W. Wells and C. Wallis. 1981. *Acropora* in Hawaii. Part 1. History of the scientific record, systematics and ecology. *Pac. Sci.* 35(1):1-14.

Irons, D., R. Kosaki, and J. Parrish 1990. Johnston Atoll Resource Survey Final Report, phase six (21 Jul 89 – 20 Jul 90). *Tech. Report* submitted to the Army Corps Engineers, Honolulu, HI. 147pp.

Jones, R. S. 1968. Ecological relationships in Hawaiian and Johnston Island Acanthuridae surgeonfishes). *Micronesica* 4:309-361.

Kopenski, R. P. and M. P. Wennekins. 1966. Circulation patterns Johnston Island, U.S. Naval Oceanographic Office, (*N00 SP-93*) Washington, DC.

Kerr, L. M. 1996. Developmental defects in damselfish (*Abudefduf sordidus*; Pomacentridae) embryos from metal artificial reefs. *Biological Bulletin* 191:306-307.

Kerr, L. M., K. Lang, & P.S. Lobel. 1997 PCB contamination relative to age for a Pacific damselfish, *Abudefduf sordidus* (Pomacentridae) *Biological Bulletin* 193:279-281

Kerr, L. M., P. S. Lobel & M. Ingoglia. 1999 Evaluation of a reporter gene system bomarker for detecting contamination in tropical marine sediments. *Biological Bulletin*, 197:79-82.

Kosaki, R. K., R. L. Pyle, J. E. Randall, and D. K. Irons. 1991 New records of fishes from Johnston Atoll with notes on biogeography. *Pacific Science* 45:186-203.

Kosaki, R. K. 1989 *Centropyge nahackyi*, a new species of angelfish from Johnston Atoll (Teleostei: Pomacanthidae). *Copeia* 1989 (no. 4): 880-886.

Lobel, P. S. 1978. Diel, lunar and seasonal periodicity in the reproductive behavior of the Pomacanthid fish, *Centropyge potteri* and some other reef fishes in Hawaii. *Pacific. Science.* 32:193-207.

Lobel, P. S. 1985. Oceanographic investigations to assess the impact of proposed deep ocean disposal of brine waste off Johnston Atoll. *Rpt. to US Army Corps Engineers*, Honolulu, HI. 135 pp.

Lobel, P. S. 1989 Ocean current variability and the spawning season of Hawaiian Reef fishes. *Environmental Biology of Fishes.* 24:161-171.

Lobel, P. S. 1997 Comparative settlement age of damselfish larvae (*Plectroglyphidodon imparippennis*, Pomacentridae) from Hawaii and Johnston Atoll. *Biological. Bulletin.* 193:281-283

Lobel, P. S. 1978 Diel, lunar and seasonal periodicity in the reproductive behavior of the pomacanthid fish, *Centropyge potteri* and some other reef fishes in Hawaii, *Pacific Science* 32: 193-207.

Lobel, P. S. 1981 Trophic biology of herbivorous reef fishes: alimentary pH and digestive strategies. *Journal of Fish Biology* 19: 365-397.

Lobel, P. S., D. M. Anderson and M. Durand-Clement. 1988 Assessment of ciguatera dinoflagellate populations: sample variability and algal substrate selection. *Biological Bulletin* 175: 94-101.

Lobel, P. S. 1988 Quantitative evaluation of brine-liquid disposal in the ocean. pp. 223-240. In: E. K. Datta (ed.) *Proceedings of the 1987 Conference on Pacific Basin Management of the 200 Nautical Mile Exclusive Economic Zone.* Pacific Basin Management Council. Honolulu, HI, July 1987.

Lobel, P. S. 1988 Biotechnology and fisheries oceanography. pp. 296-299. In: A. Vlavianos-Arvanitis (ed.) *Proceedings of the First International Conference on Biopolitics.* Biopolitics International Organization, Athens, Greece.

Lobel, P. S. 1989 Ocean current variability and the spawning season of Hawaiian reef fishes. *Environmental Biology of Fishes* 24 (3): 161-171.

Lobel, P. S. 1989 Spawning behavior of *Chaetodon multicinctus* (Chaetodontidae); pairs and intruders. *Environmental Biology of Fishes* 25 (1-3): 125-130.

Lobel P. S. & D. A. Mann. 1995 Spawning sounds of the damselfish, *Dascyllus albisella* (Pomacentridae), and relationship to male size. *Bioacoustics* 6:187-198

Lobel, P. S. 1996 Spawning sound of the Trunkfish, *Ostracion meleagris* (Ostraciidae). *Biological Bulletin* 191: 308-309

Lobel, P.S. 1997 Comparative settlement age of damselfish larvae (*Plectroglyphidodon imparipennis*, Pomacentridae)from Hawaii and Johnston Atoll. *Biological Bulletin* 193: 281-283

Lobel P. S. and L. M. Kerr, 1999 Courtship sounds of *Abudefduf sordidus* (Pomacentridae) *Biological Bulletin*, 197:18-20

Lobel P. S. and D. A. Mann, 1995 Courtship and mating sounds of Dascyllus albisella (Pomacentridae). [abstract] Bulletin of Marine Science 57(3):705

Lobel, P. S. 1998 Passive acoustic monitoring of fish reproduction coupled to physical oceanographic variables. EOS, Transactions American Geophysical Union 79(1):OS180 [abstract]

Lobel, P. S. 1998 *Coral Reef Conservation guide for the military*. Legacy Resource Management Program, US DoD, Special Publ. 12 pp.

Lobel P. S. 2001 Fish bioacoustics and behavior: passive acoustic detection and the application of a closed-circuit rebreather for field study. *Marine Technology Society Journal* 35(2)19-28

Lobel, P. S. 2002 Diversity of fish spawning sounds and the application of passive acoustic monitoring. *Bioacoustics* 12:286-289

Lobel, P. S. and A R. Robinson. 1983 Reef fishes at sea: ocean currents and the advection of larvae. *Symp. Ser. Undersea Res.* 1(1):29-38. NOAA Undersea Res. Prog., Rockville, Md.

Lobel, P. S. and A R. Robinson. 1986. Transport and entrapment of fish larvae by ocean mesoscale eddies and currents in Hawaiian waters. *Deep Sea Res.* 33(4):483-500.

Lobel, P.S., Mann, D.A. and Sancho, G. 1995. Pollution and coastal fishes: how much do we know? In: *Cleaning up our Coastal Backyards*. Ocean Policy Institute Special Publication, Center for Strategic International Studies (CSIS). Melvin Peterson ed.

Lobel P. S. and L. M. Kerr. 2003 Status of contaminants in Johnston Atoll lagoon sediments after 70 years of U.S. military occupation., In M.K. Kasim Moosa, S.Soemodihardjo, A.Nontji, A.Soegiarto, K. Romimohtarto, Sukarno and Suharsono. (Editors) *Proceedings of the Ninth International Coral Reef Symposium*, Bali, Indonesia, October 23-27 2000. Published by the Ministry of Environment, the Indonesian Institute of Sciences and the International Society for Reef Studies. Pages 861 to 866.

Lobel, P. S. & L. Kerr Lobel. 2004 Annotated checklist of the fishes of Wake Atoll. *Pacific Science* 58(1): in press

Mann, D. A. & P. S. Lobel. 1995 Passive acoustic detection of spawning sounds produced by the damselfish, *Dascyllus albisella* (Pomacentridae). *Bioacoustics* 6: 199-213

Mann D. A. and P. S. Lobel, 1995 Passive acoustic detection of fish sound production associated with courtship and spawning. [abstract] *Bulletin of Marine Science* 57(3):705-706

Mann, D.A. and P.S. Lobel 1998 Acoustic Behavior of the famselfish, *Dascyllus albisella*: behavior and geographic variation *Environmental Biology of Fishes* 51:421-428.

Mann, D.A. and P.S. Lobel1997 Information content and propagation of damselfish (Pomacentridae) courtship sounds *Journal. Acoustical Society of America* 101(6):3783-3791.

Oliver S. J. and E. Watson. 1999. Threat sensitive nest defense in domino damselfish, *Dascyllus albisella. Biological Bulletin* 197:244-246.

Oliver, S. J. 2001. Mate choice and sexual selection in the domino damselfish, *Dascyllus albisella. Boston University Marine Program Ph.D. Thesis*

Maragos, J. E. and P. L. Jokiel. 1986. Reef corals of Johnston Atoll: One of the world's most isolated reefs. *Coral Reefs* 4:141-150.

Johnson, O, D. Osborne, C. Wildgoose, W. Worstell & P. Lobel, 1997 Development and testing of a large-area underwater survey device. *IEEE Transactions on Nuclear Science* 44(3):792-798.

Ralston, S. 1981. Aspects of the reproductive biology and feeding ecology of *Chaetodon miliaris,* a Hawaiian endemic butterflyfish. *Environmental Biology of. Fishes* 6:167-176.

Randall, J. E. 1955 A revision of the surgeon fish genus Ctenochaetus, family Acanthuridae, with descriptions of five new species. *Zoologica* (N. Y.) v. 40 (pt 4, no. 15): 149-166

Randall, J. E. 1976. The endemic shore fishes of the Hawaiian Islands, Lord Howe Island and Easter Island. *Colloque Commerson,* 1973. O.R.S.T.O.M. Travaux et Documents No. 47: 49-73.

Randall, J. E. 1980. New records of fishes from the Hawaiian Island. *Pac.ific Science.* 34(3):211-232.

Randall, J.E. 1992. Endemism of fishes in Oceania. UNEP. Coastal resources and systems of the Pacific Basin: investigation and steps toward protective management. *UNEP Regional Seas Reports and Studies* No. 147.

Randall, J.E. 1995. Zoogeographic analysis of the inshore Hawaiian fish fauna. Volume I. Species systematic and information management priorities. In: *Marine and Coastal Biodivesity in the Tropical Island Pacific Region.* J.E. Maragos, M.A. Peterson, L.G. ,Eldrege, J.E. Bardach, and H.F. Takeuchi eds. *Program on Environment*, East West Center. Honolulu, HI. pp. 193-203.

Randall, J. E., P. S. Lobel and E. H. Chave. 1985. Annotated checklist of the fishes of Johnston Island. *Pacific. Science.* 39(1):24-80.

Sancho, G. 1998. Factors regulating the height of spawning ascents in trunkfishes (Ostraciidae). *Journal of Fish Biology* 53 (Supplement A): 94-103.

Sancho, G. 1998. Behavioral ecology of coral reef fishes at spawning aggregation sites. *PhD Thesis, Massachusetts Institute of Technology / Woods Hole Oceanographic Institution Joint Program in Oceanography*

Sancho, G. 2000. Predatory behaviors of *Caranx melampygus* (Carangidae) feeding on spawning reef fishes: a novel ambushing strategy. *Bulletin of Marine Science* 66 (2): 487-496

Sancho, G., Ma, D. and Lobel, P.S. 1997. Behavioral observations of an upcurrent reef colonization event by larval surgeonfish *Ctenochaetus strigosus* (Acanthuridae). *Marine Ecology Progress Series* 153: 311-315.

Sancho, G., Solow, A.R. and Lobel, P.S. 2000. Environmental influences on the diel timing of spawning in coral reef fishes. *Marine Ecology Progress Series* 206: 193-212

Sancho, G., Petersen, C.W. and Lobel, P.S. 2000. Predator-prey relations at a spawning aggregation site of coral reef fishes. *Marine Ecology Progress Series* 203: 275-288

Shelltema, R. S., I. P. Williams and P. S. Lobel, 1996 Retention and long distance dispersal between oceanic islands by planktonic larvae of benthic gastropod Mollusca_*American Malacological Bulletin* 12:67-75

Wyrtki, K. 1975. Fluctuations of the dynamic topography in the Pacific ocean. *Journal of. Physical. Oceanography* 5: 450-459.